EDITED BY
ANA JORGE AND SOFIA P. CALDEIRA

WITH AN AFTERWORD
BY PETER LUNT

ATMOSPHERES AND DIGITAL MEDIA

Connection and Disconnection Across
Everyday Life

BRISTOL
UNIVERSITY
PRESS

First published in Great Britain in 2026 by

Bristol University Press
University of Bristol
1–9 Old Park Hill
Bristol
BS2 8BB
UK
t: +44 (0)117 374 6645
e: bup-info@bristol.ac.uk

Details of international sales and distribution partners are available at bristoluniversitypress.co.uk

© Ana Jorge and Sofia P. Caldeira 2026

The digital PDF and ePub versions of this title are available open access and distributed under the terms of the Creative Commons Attribution-NonCommercial-NoDerivatives 4.0 International licence (https://creativecommons.org/licenses/by-nc-nd/4.0/) which permits reproduction and distribution for non-commercial use without further permission provided the original work is attributed.

DOI: 10.51952/9781529252644

British Library Cataloguing in Publication Data
A catalogue record for this book is available from the British Library

ISBN 978-1-5292-5262-0 paperback
ISBN 978-1-5292-5263-7 ePub
ISBN 978-1-5292-5264-4 OA PDF

The right of Ana Jorge and Sofia P. Caldeira to be identified as editors of this work has been asserted by them in accordance with the Copyright, Designs and Patents Act 1988.

All rights reserved: no part of this publication may be reproduced, stored in a retrieval system, or transmitted in any form or by any means, electronic, mechanical, photocopying, recording, or otherwise without the prior permission of Bristol University Press.

Every reasonable effort has been made to obtain permission to reproduce copyrighted material. If, however, anyone knows of an oversight, please contact the publisher.

The statements and opinions contained within this publication are solely those of the editors and contributors and not of the University of Bristol or Bristol University Press. The University of Bristol and Bristol University Press disclaim responsibility for any injury to persons or property resulting from any material published in this publication.

Bristol University Press works to counter discrimination on grounds of gender, race, disability, age and sexuality.

Cover design: blu inc
Front cover image: Stocksy/Beatrix Boros

Contents

About the Editors	iv
Notes on Contributors	v
Acknowledgements	vii
Funding	viii
Introduction: Atmospheres and Digital Media Dis/connection *Ana Jorge and Sofia P. Caldeira*	1
one — Post-Digital Parenting: The Relational-Affective Network of the Family *Francisca Porfírio, Ana Jorge and Rita Grácio*	30
two — Platformized Feminisms and Social Media Ambiences *Sofia P. Caldeira, Ana Jorge and Ana Kubrusly*	57
three — Affective Temporalities in Pilgrimage: Anticipation, Presence and (Pro)longing *Ana Jorge, Filipa Neto, Ana Kubrusly and Edna Santos*	88
four — Affective Intensities of Dis/connection in Mourning *Ionara Silva, Ana Jorge and Filipa Neto*	115
Afterword: Reflections on the Role of Atmospheres in the Mediation of Everyday Life *Peter Lunt*	142
Index	157

About the Editors

Ana Jorge (PhD) is Senior Researcher at CICANT and Associate Professor at Lusófona University. Ana works in media and cultural studies, particularly researching audiences, celebrity and influencer culture, digital culture, children/youth and families. She led the funded projects Dis/Connect (2021–22) and On&Off (2023–24). Ana's scholarship appears in journals such as *New Media & Society*, *Social Media and Society*, and *Information, Communication & Society*. She is the co-editor of *Digital Parenting* (2018), *Reckoning with Social Media* (2021) and *Audience Interactions in Contemporary Celebrity Culture* (2024).

Sofia P. Caldeira (PhD) is a Researcher at CICANT and Assistant Professor at Lusófona University. She holds a Communication Sciences PhD from Ghent University, Belgium (2020), funded by the Fundação para a Ciência e a Tecnologia, and is a former Marie Skłodowska-Curie fellow. Her research focuses on digital cultures, with a particular focus on social media and Instagram, exploring politics of gender representation and everyday politics in these digital contexts. She also currently serves as Chair of ECREA's Digital Culture and Communication section.

Notes on Contributors

Rita Grácio (PhD) is Assistant Professor at Lusófona University. As an integrated researcher at CICANT-Lusófona University, she takes part in several European-funded research projects. Grácio's research focuses on cultural and creative industries, audience studies, cultural production, gender studies, and digital culture/digital communication.

Ana Kubrusly is a PhD candidate in Communication Sciences at NOVA University of Lisbon, funded by the Portuguese Foundation for Science and Technology (https://doi.org/10.54499/PRT/BD/154767/2023). Her research focuses on children's and adolescents' relationships with digital environments regarding their literacy, skills, well-being and digital cultures.

Peter Lunt is Professor of Media and Communication at the University of Leicester, UK. His research has focused on the media audience, media regulation, and media and social theory. In addition to over 100 articles and six books, Peter's recent publications include a monograph for Polity titled *Goffman and the Media* and an edited collection seeking to capture the emerging agenda for audience research, *The Routledge Companion to Media Audiences* (co-edited with Annette Hill).

Filipa Neto is a PhD candidate in Communication Sciences at Lusófona University and CICANT, with funding from Fundação para a Ciência e a Tecnologia (grant 2024.01862. BD). Her research is centred on the themes of well-being, digital media and spirituality. She holds a MA in Sociology at ISCTE-IUL and a BA in Social and Cultural Communication from the Catholic University of Portugal.

Francisca Porfírio is a Communication Sciences PhD student at Lusófona University and CICANT, with an individual grant from Fundação para a Ciência e a Tecnologia (grant 2021.07777.BD). Francisca has a MA in Communication Sciences from the Catholic University of Portugal and a BA in Sociology from ISCTE-IUL. She works on the topics of digital parenting and social media.

Edna Santos obtained her MA in Organizational Communication at Lusófona University in 2025. Her research focused on how digitalization was perceived by pilgrims as well as touristic agents in Camino de Santiago.

Ionara Silva obtained her PhD in Communication Sciences at Lusófona University and CICANT in 2025. Her dissertation explored the intersection between mourning, death and digital technologies.

Acknowledgements

We thank André Jansson and Aleena Chia for their generous support as the project's advisers, from the start to critical moments of the project's development.

We would also like to acknowledge Ranjana Das, Francesca Pasquali, Emma Beuckels, Ludmila Lupinacci, Sarit Navon, Mats Nilsson, Magdalena Kania-Lundholm and Karin Fast for commenting on previous versions of the chapters in this book. We appreciate the kind words of encouragement and wise leads to improve our work.

The team acknowledges Ana Marta Flores for her generous guidance and support in data scraping and preparation.

Any project is a combination of ideas and logistics, and we are very grateful to CICANT's staff members Margarida Santos and Ana Parruca for all the bureaucratic and administrative support.

Funding

The research reported in this book was produced under the grant by Fundação para a Ciência e a Tecnologia, IP, with grant 2022.01282.PTDC (https://doi.org/10.54499/2022.01282.PTDC). The publication with Open Access received support from CICANT research unit, from Lusófona University, nationally funded by Fundação para a Ciência e a Tecnologia, IP (https://doi.org/10.54499/UIDB/05260/2020).

We would also like to acknowledge the European Union's Horizon Europe funding received by Sofia P. Caldeira under the Marie Skłodowska-Curie grant agreement No 101059460.

Introduction: Atmospheres and Digital Media Dis/connection

Ana Jorge and Sofia P. Caldeira

As we wrapped up this book in April 2025, a massive power outage affected the Iberian Peninsula for about 12 hours (Jolly, 2025). First, the electricity went out, and then all connectivity. This unprecedented failure in infrastructures and networks prompted emergency plans from governments, as well as adjustment and coping from individuals with no phone reception or internet access. In the days and weeks after the incident, reactions on mainstream media opinion pieces and takes on social media from ordinary people seemed to be divided. On the one hand, we could see stories of those embracing being unplugged and unreachable for quotidian obligations, engaging with others in the street if they happened to be caught in the cities' centres and away from home, or patiently letting time take its course and waiting for normal life to be resumed. Mindfulness tropes were activated (Baym et al, 2020). Here, 'good old' radio was celebrated for its capacity to keep people informed throughout an emergency. On the other hand, we found reports of anxiety for having felt out of control, especially among people with caregiving responsibilities. This impressionist – and simplistic – portrayal illustrates how people seem to resent being under pressure to be online and connected, as much as they do having tech failing them (Paasonen, 2015). However, it is not a matter of identifying which group one might fit, but how these positions alternate or even accumulate.

What this disruption showed so clearly is how deeply reliant on and entangled with digitalization our societies are. It has become commonplace to observe the current panorama of media use in everyday life as one of media ubiquity but also intrusiveness (Mollen and Dhaenens, 2018), and increasingly so in highly mediatized and datafied societies (Couldry and Hepp, 2017). For ordinary individuals, in their capacities as family members, workers, consumers or citizens, this pervasiveness represents expectations of 'always-on' availability, and an abundance of media content constantly at hand (Boczkowski, 2021). For professionals such as journalists, there is growing pressure to continually present themselves in ways that erode the limits between personal and professional spheres (Bossio and Holton, 2021). There are superabundant options for micro-sociality, made to facilitate interaction online but eventually turning into burdensome and felt as sticky (Brubaker, 2023): think of the read receipts or the '[someone] is typing' feature on WhatsApp or other instant messaging services, and how they keep us on hold for further communication and press us for a response. Gestures such as swiping left or right, or the pull to refresh symbolize our engagement with the realtimeness orchestrated by platforms and devices (Weltevrede et al, 2014; Lupinacci, 2024).

Digital media are said to make individuals 'dependent, distracted, bored' (Paasonen, 2021). Social media platforms are particularly criticized for inducing addictiveness through their design, deploying unfair business practices, undermining democracy through disinformation (Chia et al, 2021), and aggravating a mental health crisis by fostering bodily and social comparison. But, as part of the everyday, digital and social media are also sources of and means for *captivation* (Coleman, 2025), joy, pleasure, fun and excitement, comfort and safety (Lupinacci, 2021; Lehto and Paasonen, 2021; Bengtsson and Johansson, 2022). With and through digital media, people go through rituals of getting up and going to bed, commuting, spend evenings

INTRODUCTION

'Netflix and chilling', laugh and reflect with memes, find others who think the same, or vicariously experience things they wish they were living, receive recognition and validation, cherish memories with significant others. As both the case of the power outage and these diverse, and at times conflicted, accounts illustrate, encounters with digital and social media can change and drift in response to particular affective atmospheres, which are both individually and collectively embodied, and affected by spatial, material and cultural circumstances (Riedel, 2019).

This ambivalence leads individuals to (try to) 'disentangle' from the digital (Jansson and Adams, 2021), 'reckon' with social media (Chia et al, 2021) or 'opt out' (Brennen, 2019) in temporary and partial, more often than radical, forms. While some argue that there is 'nothing to disconnect' from (Bucher, 2020) and that 'disconnection is futile' (Lomborg, 2020), as datafication is all-encompassing and co-opts even the acts of disconnecting, we argue there is value in further illuminating practices and meanings of engaging with and disengaging from digital media in everyday life in different settings. This can contribute, we believe, to better understanding the circulating narratives and discourses surrounding dis/connectivity and the role of digital media in social life (Syvertsen, 2020; Paasonen, 2021) and ultimately to ground critical discussions on the implications of such processes.

Our proposal is to make sense of dis/connection practices in a practice-based approach, as part of everyday life, through the notion of *atmospheres*. This Introduction aims to set the scene for the following chapters by recalling the growing scholarship on digital disconnection and an emergent interest in affect and atmospheres as a framework in audience research. We then position our vision within these frameworks and their usefulness in understanding dis/connection, and present the research project from which we draw the results that underpin this book. Lastly, we outline the book structure with a summary of the contributions.

Dis/connecting

The research that forms the basis for this book stems from engagement primarily with literature on digital disconnection. Research on how people disconnect from one or more forms of digital media has expanded and consolidated over the last decade, bringing together perspectives from different disciplines, from media studies to computer science and human-computer interaction, psychology, cultural studies and more. Regardless of whether it is a field in its own right, a sub-field or a corner, this fast-growing body of literature emerged as a generative 'entry point into a wider reflection on sociality, agency, rights and everyday life' (Moe and Madsen, 2021, p 1587) as well as 'identity and self ..., inequality, labor, ... and political engagement' (Lomborg and Ytre-Arne, 2021, p 1532) in a world permeated, and increasingly saturated, by digital media, as we alluded to earlier.

We regard digital dis/connection as 'embedded in the everyday lives of individuals, encompassing their negotiations of online and offline uses in place, time, and social context, and thus [as] a fluid, ambivalent, highly contingent, and reversible process' (Jorge, 2019, p 3). By now, there is a consensus that radical or definitive disconnection is unattainable, and thus, that it is best seen as inserted in a continuum of practices (Ribak and Rosenthal, 2015). Thus, the slash in 'dis/connecting' conveys how both acts are inherently linked. In popular and academic vocabularies, terms such as 'withdrawal, opting out, leaving, non-use, non-participation, detox, unplugging' (Kuntsman and Miyake, 2019, p 906), but also avoidance, refusal, interruption, or abstention are used to refer to disconnection (for example, Portwood-Stacer, 2013; Jorge, 2019; Aharoni et al, 2021). A comprehensive conceptualization of disconnection practices (Light, 2014) helps discuss different levels of activity, while acknowledging them all as meaningful. This means considering technical, personal and political tactics together (Plaut, 2015;

Mannell, 2019), as well as individual and collective forms of disconnection (Kaun and Treré, 2020).

Scholarship on disconnection has paid particular attention to online communication, particularly social media and mobile devices (notably smartphones; Mannell, 2019; Ytre-Arne et al, 2020), streaming and mobile connectivity. The fluidity created by the ubiquity and constancy of access through mobile media poses challenges to navigating conflicting demands and goals. People's perceptions shift between seeing the digital media as *saving* or *stealing* time, facilitating everyday micro-coordination *and* taking away opportunities for face-to-face, *genuine* connection (Ytre-Arne et al, 2020; Syvertsen and Ytre-Arne, 2021), or helping navigation and affording safety *while* making it challenging to separate holidays from routine life (Rosenberg, 2019).

Another axis of disconnection scholarship has been individual and voluntary acts of disconnection (Bagger, 2024). This scholarship derives from the interest in analysing market initiatives such as detox camps as well as media and popular self-help literature (Syvertsen and Enli, 2020). Across this work, disconnection is read as an ideological (neoliberal) construction that transfers responsibility onto the individual, with the promise of its instrumental capacity to increase their well-being, without sufficient reform of digital media's (infra)structure. Examples in this case would be the popularization of not just 'digital detox', but also vernacular expressions such as digital 'Sabbath', 'hygiene', 'cleanse' or 'declutter' (Syvertsen, 2020). Partially in overlap with this axis, disconnection studies have focused on acts undertaken by privileged sections of society, along lines of class as well as gender (Beattie, 2020), and mostly originated from Global Northern perspectives (Bagger, 2024).

Nevertheless, increasing attention has been paid to nuanced aspects of disconnection and particularly social inequalities within and across societies (Mutsvairo et al, 2023; Bozan and Treré, 2024). This effort represented a focus (back) on material aspects of connectivity and access, as well as a problematization beyond voluntary disconnection into involuntary and unwanted forms of disconnection (even without political intent as in the shutdowns catalogued by Kaun and Treré, 2020). Pype (2021), for example, looks at the ordinary experiences of 'unexpected and undesired disconnection' in Kinshasa, Democratic Republic of the Congo, in the form of 'involuntary logging off of the Internet due to lack of battery time or credit time' (p 1200). Disconnection, in this case, is very different from the breaks in networks and services experienced by Northern European individuals (Paasonen, 2015) – or in Portugal and Spain in April 2025.

This field has grown in tandem with the dislike and disappointment with social media for failing to live up to its democratizing potential (Syvertsen, 2017) and a wide backlash towards the platform economy (Chia et al, 2021; Albris et al, 2024). Furthermore, this critique has evolved during and after the pandemic expansion of connectivity (Jansson and Adams, 2021; Brubaker, 2023). Disconnection studies have contributed to wider critiques of the role of platforms in a neoliberal economic framework, often emphasizing how disconnection practices sustain rather than transform the system (Light, 2014; Karppi, 2018; Jorge, 2019).

In fact, in many ways, platforms and the market have incorporated or co-opted disconnection in response to backlash. On the one hand, platforms have developed a variety of 'features for disconnectivity' to allow users to create distance from others and modulate connectivity (Mannell, 2019; John and Katz, 2023) and 'screen time' functions to manage the time spent on apps (Beattie and Daubs, 2020; Jorge et al, 2022). Ultimately, these features and functions can be seen as 'lubricant' mechanisms that help retain users on the platforms with a qualified engagement, therefore avoiding that they leave completely (Light and Cassidy, 2014), and restore the platforms' public image. On the other hand, the market has further exploited the desires to manage connectivity or to conspicuously appear disconnected. Here, one can find apps to control disconnection to increase focus, such as Forest, objects such as 'the phone box' (Fast and Syvertsen, 2024), or the popularization of analogic and retro design in digital objects (Thorén et al, 2019).

Our book particularly draws on a strand of scholarship on dis/connection rising within cultural studies and audience studies, which has examined the affective dimensions of connectivity and disconnection (Karppi, 2018; Paasonen, 2021; Bossio and Holton, 2021; Lehto and Paasonen, 2021). This group of work emphasizes feelings and sensations that lead to disconnecting, such as disenchantment, boredom, overwhelm or emotional fatigue (Jorge et al, 2024); and/or those that result from disconnecting, such as ambivalence or anxiety (Chia et al, 2021). Such perspectives shift the focus 'from ends to the means of disconnection' (Karppi et al, 2021, p 1599). Here, we find work considering the sensations afforded by commodities specifically designed to disconnect (Chia and Beattie, 2021); services that promise to support individuals to disconnect from digital media or invite them to imagine disconnecting (Karppi et al, 2021; Jorge, 2024); the affective reactions to unforeseen and uncontrollable aspects of digital life (Maslen, 2022) or to unexpected disconnection via breaks and glitches (Paasonen, 2015). This brings us to the rise of affect in audience studies, and to everyday life and media.

Affect and the everyday

Several scholars have noted the affective turn in cultural studies since the 1990s, and progressively in media and communications, particularly in audience studies. In 2011, Gibbs observed that 'affect theory has not yet been seriously adopted at all by audience research' (Gibbs, 2011, p 252). Gibbs asserted that adopting affect theory would enable audience research to capture the *flows* between subjects and objects more effectively, transcending stark divisions between the individual and the social. More recently, Hermes (2024) writes that earlier audience research associated with cultural studies focused on 'meaning making, pleasure and ideology' (p 350) in everyday and mundane practices with media, but that affect emerged as 'a prominent theme [which] becomes clear when looking at the change from broadcast media to platformised, online media' (p 352). This attention to affect can be traced to Papacharissi's influential work on social media (2014) and Wahl-Jorgensen's (2019) on news; and to the influence of queer and gender studies, bringing in questions of 'embodiment, energy and belonging' (Hermes, 2024, p 352), dispositions, resonance and contagion (Gibbs, 2011).

The terms 'affect' and 'emotion' are sometimes used interchangeably, yet it is essential to distinguish them (Bericat, 2016). Affect is understood to precede, to 'usher in, produce and give rise' to emotions (Tucker and Goodings, 2017, pp 629–630). While affect is diffuse and ethereal, it is detectable as 'affective states' (Petit, 2015). Affect is complex and complicated, as it collapses 'somatic, neural, phenomenological, discursive, relational, cultural, economic, developmental, and historical patterns [that] interrupt, cancel, contradict, modulate, build and interweave with each other' (Wetherell, 2012, p 14). Under this universe, we can also find *moods*, understood as more lasting affective states, without a specific target and even if feeble, and potentially contagious to others (Bericat, 2016).

In turn, emotions are considered more personal and individual; they are situated and can be 'identifiable states grounded in our embodied life histories' (Ahmed, in Paasonen, 2016). Emotions can take up organized, collective shapes in terms of sensations and cultural meanings (for example, the pessimism of an era). Media and culture are important agents in the socialization of emotions (Williams, 1977; Illouz, 2008), in encoding not just how one might feel in particular situations but also what are the appropriate ways to express those emotions – and by whom.

It thus follows that media and audience but also digital media studies can find in affect a generative lens to discuss engagement

INTRODUCTION

with and through media, making sense of 'repetitions, pains and pleasures, feelings and memories' (Wetherell, 2012, p 2). Through affect, psychological and physiological, as well as social aspects of engagement with media are considered (Dahlgren and Hill, 2020).

Particularly, engaging with affect facilitates unpacking routines as sites of power (Koivunen et al, 2024), as it allows one to pay attention to the microscopic, infinitesimal scales of actions related to the media. Routines and rituals, the quotidian and the 'taken for granted', form *everyday life*, encompassing diverse social experiences (Silverstone, 1994). The everyday is both omnipresent and invisible, public and private, commonplace and capable of transcendence (Lefebvre, 1991). It follows particular cycles and rhythms (Lefebvre, 2004) and evokes experiences of both comfort and boredom. Yet, everyday life can also be understood as a process, inherently open to change and the setting where new knowledge can be encountered and new habits developed (Highmore, 2011). Everyday life studies have drawn on diverse theoretical and empirical lenses that highlight their proneness to ambiguity (Highmore, 2002). Media and everyday life has been explored particularly within audience studies (for example, Silverstone, 1994; Hermes, 1995; or, more recently, Ytre-Arne, 2023). Media, digital or otherwise, has long become a part of the habitual, tacit, or unspoken sensitivities of everyday life (Pink and Leder Mackley, 2013), both in the domestic realm (Highmore, 2011) and, with the rise of mobile media, increasingly outside the home too (Ytre-Arne, 2023). Media like television (Silverstone, 1994) and, more recently, social media platforms have become a taken-for-granted presence in homes, cafés and other mundane settings, becoming part of daily conversations (Lupinacci, 2021; Bengtsson and Johansson, 2022). However, even though the media has 'saturated' the everyday, people's lives are not necessarily 'media-centric' (Couldry, 2012).

In this book, we explore media engagement – and disengagement – alongside people's social practices or affective experiences in a practice-centred approach. As such, it is crucial not only to foreground people's lived experiences on and with digital media, the everyday processes of meaning-making surrounding these practices, and the ambivalent affective experiences they evoke, but also to recognize the bi-directional influences in how digital media and everyday life (re-)shape one another (Ytre-Arne, 2023). We present our conceptual approach in the next section.

Atmospheres in media dis/connection: a conceptual approach

Under this affective turn, we can find the growing interest in 'atmosphere' as a framework. The term atmosphere 'has drifted from physics into being a description of emotional moods or situations: from the original meaning of a sphere of gas surrounding a body such as a planet, to a prevailing psychological climate' (Ehn et al, 2016, p 83). Some elements are retained in characterizing atmospheres emotionally: the pressure, weight, temperature, humidity; but also, other sensorial aspects such as soundscapes and 'smellscapes' (Ehn et al, 2016).

In the 1960s, Schmitz introduced the concept of atmosphere as part of his 'non-cognitive, social phenomenology of emotions as these play out in interpersonal space through the felt body' (Lunt, 2024, p 473). Böhme's take on atmospheric architectures as 'felt spaces' (Riedel, 2019) is also highly influential to this framework. Atmospheres are both staged and completed by those participating in them, therefore dynamic and not pre-determined. According to geographer Ben Anderson (2009, p 80), 'affective atmospheres' are 'singular affective qualities that emanate from but exceed the assembling of bodies'. Materiality and intersubjectivity are amalgamated. Atmosphere 'construes feelings as collectively embodied, spatially extended, material and culturally inflected' (Riedel, 2019, p 86). Shared across disciplines is the vision of atmospheres as between, or beyond, distinctions of inner and outer world, individual and collective, body and mind.

Using the concept of atmospheres draws our analytical attention to how 'moods are produced, sustained, or changed', but also 'how people come to share an atmosphere or are taken in by it and how it may dissolve boundaries not only between people but also between the body and the material surroundings' (Ehn et al, 2016, p 83). In this sense, this approach decentres the human being, perceiving interactions and movements of human and non-human elements in specific settings as co-constitutive (Lupton, 2017; Sumartojo and Pink, 2018).

For audience research, this means going beyond perception and interpretation into broader notions of engagement and attunement (Lunt, 2024), decentring analysis from text and discourse, and paying more attention to material and sensorial aspects (Hillis et al, 2015). 'Atmosphere' captures global and social flows as well as individual feelings. Moreover, as in meteorology, an atmosphere is potential – it enables and makes possible, including what is imagined – as much as it is about containment. This perspective can illuminate 'what media texts and platforms afford and close off' (Hermes, 2024, p 356). It is also important to emphasize that not everyone feels the atmosphere in the same way nor has equal power to affect it (Griffero, 2021; Kolehmainen and Mäkinen, 2021), for example parents and children, which meets old questions in audience research concerning power relations at micro-scales.

In our exploration of the framework to make sense of people's dis/connection with digital media, 'affective atmospheres' is a fruitful approach to understand digital media as part of everyday life, and make sense of the encounters between human elements and devices, services or platforms, in situated contexts. Besides considering digital media as objects or devices and as content or text, this framework also acknowledges the role of symbols and ideologies surrounding these non-human elements, which are formed socially and personally (Illouz, 2008). If we understand those encounters as interactions where one part does not command the other, we can overcome a notion of media as 'tools' one 'uses' and that 'afford' something to the user, and thus break with the very discourses that disconnection research has denounced: for example digital media as tools the user should be able to control. Tim Markham (2020, p 25) rightly poses: '[W]e should not think of selves who go out into the world and use digital technologies; they are always already part of the world in which we are forever emerging, in the countless acts of subjectivation through which we make that world and ourselves familiar and navigable.'

The notion of atmospheres considers the dynamic interplay of moods and sensations, which are not always conscious, and how this moves people towards and/or away from digital media. Acknowledging the 'rhythms, valences, moods, sensations, tempos, and lifespans' (Stewart, 2011, p 445) of atmospheres makes it possible to capture the oscillation and vagueness in engagement with digital media, where conflicting desires coexist, as potentiality rather than ambivalence. Considering how personal dispositions and the material conditions of everyday life motivate how people turn to and away from digital media, and also the ways in which these come to affect how people feel, offers a better understanding of how disconnection may emerge, and in what shapes. This might also escape normative views on what disconnection might bring, and focus on how disconnection is 'embodied and affectively experienced' (Coleman and Paasonen, 2020, p 1).

Furthermore, our book endeavours to extend the framework of affective atmospheres to the articulation of the everyday digital media *and* in-person as one environment, seeking to capture the situatedness of media engagement in sociocultural context, as it is co-produced and where contagion can occur (Sampson et al, 2018; Kolehmainen and Mäkinen, 2021). At a time when 'vibes' have become not just vernacular(ized) (Salazar, 2023) but also co-opted by platforms (Lupinacci, 2025) and marketing, we wish to contribute to further enrich the territory intersecting affect and digital media (Hillis et al, 2015; Sampson et al, 2018). We investigate this by focusing on micro-environments across everyday, social situations, heuristically chosen to explore differences in circumstances where disconnection might appear out of distress or in elective situations. Here, we take a broad understanding of the everyday, embracing both mundane settings and the reactions to moments of disruption that permeate them.

Project On&Off: methodological and ethical approach

This book reports on extensive empirical research conducted under the project *On&Off: atmospheres of dis/connection*, which ran between 2022 and 2024, funded by the Portuguese Fundação para a Ciência e a Tecnologia. In this project – and this book – we address different dimensions of affective atmospheres in media dis/connection by bringing together four case studies exploring different everyday settings, where mundane experiences can concur with moments of affective disruption: family life, activisms, pilgrimages and mourning. Across these experiences, we seek to explore the shifting and often unconscious affective engagements and disengagements with digital media in everyday life.

Our empirical research centres on the national context of Portugal. Portugal occupies an interesting position at the periphery of Europe, lagging behind the economic development of central and northern European countries. In a country marked by stark inequalities, its digitalization processes have been slower than the European average and persisting indicators of digital divide, chiefly among the country's older populations (Gomes and Dias, 2025), coexist with a growing internet penetration rate, estimated at 89 per cent in early 2025 (DataReportal, 2025). In this context, engagement with digital and social media, particularly in mobile devices, has become part of the quotidian routines of many Portuguese people, intersecting with different personal and sociocultural experiences (Amaral et al, 2023). National digital practices are embedded in global communities of practice, drawing on broader conventions and strategies, as studies of social media influencer cultures exemplify (Jorge et al, 2018). However, national historical, political and cultural specificities of a just 50-year-old democracy also help to shape digital uses, as we can observe, for example, in the tensions generated by a conservative and Catholic national tradition in digital practices of increasingly secular pilgrimage (Jorge, 2023) or feminist activism (Caldeira and Machado, 2023).

As pointed out earlier, disconnection scholarship is only gradually including perspectives from the Global South and disenfranchised portions of society. As such, this book contributes to expanding and decentring scholarly discussions that often privilege Anglo-Saxon or northern European contexts. Although centred in the Portuguese context, our case studies foster reflections that can be significant at wider transnational scales. The quotidian digital media uses observed, for example, among Portuguese families or feminist activists respond to local specificities while considerably drawing from global social media cultures of use. Transnational links also emerge more clearly in the cross-boundary mobility and global community-building experienced by pilgrims; or in the mourning experiences of participants who, by their migrant status, contend with ambivalent feelings of proximity and distance. Our studies thus include a diversity of participants and experiences within this national context, broadening its insights.

Empirically engaging with multifaceted affective atmospheres surrounding everyday dis/engagements with digital media raises pressing methodological questions. How can we holistically engage with people's everyday experiences with and of media, recognizing their interactions between human and non-human elements? How can we consider not only lived experiences but also the ways these intersect with imagination or memories? How can we take into account the often elusive affects these experiences evoke?

INTRODUCTION

To engage with the confluence of digital media dis/engagement, this book is grounded on an ethnographic approach – long recognized as a critical means of engaging with the complexities and subtleties of everyday relationships with media (Ang, 2006; Pink et al, 2017). As our four case studies foreground situations characterized by affectivity, vulnerability and liminality, this deep ethnographic engagement is key to grasping the affective and sensorial complexities involved in family dynamics, activist contexts, pilgrimage settings, or mourning experiences. The four case studies share the methodological basis, combining data collected through in-depth interviews, as well as digital and offline observations. Each case, however, has its specificities that call for particular methodological decisions – the details of which are available in each of the corresponding chapters.

This ethnographic combination of interviews with on- and offline observations of relevant (digital) spaces, events and communities allowed us to engage with digital media's atmospheres at various levels. As a qualitative and relational method, in-depth interviews can facilitate situated affective encounters (Ayata et al, 2019), allowing us not only to inquire about media experiences and practices of meaning-making surrounding them, but also to be attentive to non-linguistic dimensions that can reveal different affective intensities, including silences or the emergence of ambivalent feelings. However, as explored before, affect has a pre-cognitive dimension, often being experienced as vague forces or intensities that are not easily put into words (Frykman and Povrzanovic Frykman, 2016; Kennedy, 2018). While we recognize that this can pose a limitation to our approach, by complementing interviews with on- and offline observations, we open space to perceive how affect can circulate in both technologically mediated forms and in-person settings, in fleeting and multilayered micro-events (for example, Ehn et al, 2016; Nau et al, 2023). These ethnographic observations allow immersion in the participants' lifeworld, being holistically attentive to the ambiguities of affective experience and those 'indefinite somethings' that shape everyday life (Frykman and Povrzanovic Frykman, 2016). This approach also eschews strict distinctions between connection and disconnection, foregrounding how these dynamics shift in intensity, or in response to situational changes.

This book thus draws on an extensive and diverse empirical dataset, including over 90 interviews, more than 7,500 digital objects (including posts, videos and ephemeral content collected across a range of social media platforms like Instagram, Facebook and TikTok), and extensive fieldnotes from ethnographic observations of family homes, feminist protests, pilgrimage sites and bereavement groups. Fieldwork was conducted in 2023 after a hard lockdown policy in 2020–21. By 2023 the acceleration of digital hyperconnectivity brought about by the COVID-19 pandemic (Brubaker, 2023) had, for many, become normalized into a seemingly unavoidable part of quotidian experience, causing both the benefits of and the mounting criticisms against pervasive digital media to enter everyday discourse.

We adopted a situational approach to research ethics, broadly inspired by the recommendations of the Association of Internet Researchers (franzke et al, 2020). While the specificities of each case study call for particular ethical decision-making (see the following chapters for details), our approach shares an overall commitment to respecting users' expectations of visibility, ethically engaging with publicly available digital data whenever possible, adopting strategies for informed consent when dealing with private or semi-private data or directly engaging with participants, anonymizing identifiable attributes (unless requested otherwise by participants), and practising secure data storage and management.

Overview of the book

The book presents refined versions of papers presented at a one-day workshop in Lisbon and online, on 24 October 2024.

INTRODUCTION

Each empirical chapter foregrounds a different dimension of affective atmospheres in relation to each of the case studies in On&Off, with a heuristic aim to demonstrate the ductile applicability of the framework of atmospheres. Namely, we explore: the affective relationships of parents and children with and around digital media; the interplay between affective states and social media ambiences among activists; the affective temporalities in pilgrimage as they are constructed by communities as well as the platforms; and the intensities at play in using, not using, or modulating digital media during grief processes. A few themes recur across the different chapters: embodiment, affective states and moods (Chapters 2 and 4); space and ambience (Chapters 2 and 3); temporality (Chapters 3 and 4); intersubjectivity (especially Chapter 1); and norms and morality.

In Chapter 1, Francisca Porfírio, Ana Jorge and Rita Grácio explore contemporary dynamics of 'doing family' in the highly digitized and mediatized contemporary world, where home life and interpersonal family dynamics are permeated by pressures for constant connectivity. Drawing on empirical research of nine parents' social media sharenting practices and ethnographic observations in five family homes, this chapter highlights the relational and affective labour required of both parents and children to deal with digital media in everyday life. If families recognize and value digital media to cement social ties or entertain (or pacify) children, they also acknowledge the challenging feelings these media evoke, emerging from pressures and anxieties to be constantly available, normative valuations of unmediated family time, or conflicts arising from children testing limits of dis/connectivity rules imposed by parents. Managing the wide range of practices that fall under the umbrella of digital parenting, as well as the pressures that arise from them, thus demands not only digital skills but also relational and affective labour, which is often gendered and invisible.

Chapter 2, by Sofia P. Caldeira, Ana Jorge and Ana Kubrusly, explores the affective impacts of platformized feminisms for both extraordinary moments of protest and everyday encounters with feminist content. This chapter departs from the study of 2023's International Women's Day march, combining on- and offline observations and interviews with feminist activists to explore the tensions and ambivalences contained in experiences of platformized feminisms. Despite the enduring symbolic and political significance of in-person mobilizations, the use of social media for feminist practices has become increasingly naturalized – for calls for action and building anticipation for offline events, feminist community-building, or sustaining longer-term feminist engagements. These digital feminist practices are shaped by the interplay between activists' affective states and social media ambiences, shifting according to users' transient moods and feelings, the socioeconomic and technological characteristics of platforms, and the communities built around them, or even fuzzier platform vibes.

In Chapter 3, Ana Jorge, Filipa Neto, Ana Kubrusly and Edna Santos engage with experiences of pilgrimage in a post-secular context where ideas of religion, spirituality, self-improvement and tourism intermingle. Pilgrimages both suspend everyday life and permeate it. Grounded on extensive empirical research on three sites of pilgrimage – Santiago de Compostela, Fátima and the 2023 World Youth Day in Lisbon – this chapter brings together digital observations in key platforms like Instagram, Facebook, TikTok, interviews with people involved with pilgrimages, and ethnographic explorations of pilgrimage routes. The chapter foregrounds how digital media atmospheres interact with, shape, and can be conducive to different affective experiences fluctuating through time. Chapter 3 explores how different moments of pilgrimage – before, during and after these events – evoke, build and augment collective feelings of anticipation, presence and (pro)longing that suffuse social media atmospheres. Crucially, the analysis highlights how contradictory affects circulate at each of those moments. These practices emerge as cyclical in nature, eschewing the clear-cut divisions that pervade disconnection rhetoric.

Finally, in Chapter 4, Ionara Silva, Ana Jorge and Filipa Neto explore how relationships with digital and social media can shape experiences of grief and mourning. Drawing on a two-part study of mourning that includes the analysis of nine bereavement Facebook groups, and interviews with both moderators of bereavement support groups and people in mourning, this chapter addresses mourning as simultaneously a major disruption that alters quotidian life and digital routines, and as an affective force that seeps into everyday life. This chapter highlights the affective intensities in the interplay between different feelings throughout the grief processes and the ambiguities and tensions brought about by the use of digital media and the efforts to contain them. Digital and social media can be important repositories for memories of deceased loved ones and spaces to build community and seek support from others who share similar experiences. However, conflicts and discomforts caused by experiences of online mourning can create an ambivalent pull for dis/connection. Experiences of mourning can evoke conflicting desires to avoid the emotional overload of digital media, leading mourners to negotiate fears of oversharing, avoiding too intimate or prolonged expressions of grief, or shifting their presence to narrower digital spaces and sub-platforms, such as topical Facebook groups, in order to avoid clashing with dominant cultures of use that frame social media as 'cheerful' and entertaining spaces. Here, dis/connection appears as a way to deal with intense and often difficult to articulate experiences of grief, and the emotional fatigue they can carry.

The Afterword written by Peter Lunt brings the book to a close with a critical reflection on how the atmospheres generated by media – both by design and through subjective aspects – are affected by felt experiences of mediation of everyday social practices. In this Afterword, Lunt starts by offering a review of the preceding chapters, identifying the key concepts that emerge in this work and considering how each case study reflects on the embedding of social, cultural and audiences in a media-saturated world. Media, in these contexts, emerges not simply as constituting a space of and for experience, but also providing complex formations of affordances and resources that permeate everyday social routines and relationships. The Afterword highlights the central idea, running throughout the book, of the role of emotions in the mediation of everyday life, in which the bodily and affective aspects shape the experience and participation in mediated affective atmospheres. It foregrounds the prelinguistic, affective and social phenomenology of audience engagement with media, as well as the affective tensions these can encompass. Lunt's Afterword puts the work developed in this book in conversation with various theories and concepts within the fields of media and cultural studies, recognizing the underlying sociological aspects of the case studies and the role of the cultural politics of emotion to structure the studied affective atmospheres. In this way, this concluding chapter opens up questions for continued exploration of affective atmospheres in and across complex everyday experiences of digital media.

References

Aharoni, T., Kligler-Vilenchik, N. and Tenenboim-Weinblatt, K. (2021) '"Be less of a slave to the news": a texto-material perspective on news avoidance among young adults', *Journalism Studies*, 22(1), pp 42–59.

Albris, K., Fast, K., Karlsen, F., Kaun, A., Lomborg, S. and Syvertsen, T. (2024) *The Digital Backlash and the Paradoxes of Disconnection*. Nordicom, University of Gothenburg. Available at: https://urn.kb.se/resolve?urn=urn:nbn:se:norden:org:diva-13243

Amaral, I., Antunes, E. and Flores, A. M. (2023) 'How do Portuguese young adults engage and use m-apps in daily life? An online questionnaire survey', *Observatorio (OBS*)*, 17(2). https://doi.org/10.15847/obsobs17220232141

INTRODUCTION

Anderson, B. (2009) 'Affective atmospheres', *Emotion, Space and Society*, 2(2), pp 77–81.

Ang, I. (2006) 'On the politics of empirical audience research', in Durham, M. G. and Kellner, D. M. (eds) *Media and Cultural Studies: Key Works*. Blackwell Publishing, pp 174–194.

Ayata, B., Harders, C., Özkaya, D. and Wahba, D. (2019) 'Interviews as situated affective encounters: a relational and processual approach for empirical research on affect, emotion and politics', in Kahl, A. (ed) *Analyzing Affective Societies: Methods and Methodologies*. Routledge, pp 63–77.

Bagger, C. (2024) 'A decade of digital disconnection research in review: where, what, how, and who?', in Albris, K., Fast, K., Karlsen, F., Kaun, A., Lomborg, S. and Syvertsen, T. (eds) *The Digital Backlash and the Paradoxes of Disconnection*. Nordicom, University of Gothenburg, pp 109–128. https://urn.kb.se/resolve?urn=urn:nbn:se:norden:org:diva-13243

Baym, N. K., Wagman, K. B. and Persaud, C. J. (2020) 'Mindfully scrolling: rethinking Facebook after time deactivated', *Social Media + Society*, 6(2). https://doi.org/10.1177/2056305120919105

Beattie, A. (2020) The manufacture of disconnection. Doctoral thesis (PhD), Te Herenga Waka–Victoria University of Wellington.

Beattie, A. and Daubs, M. S. (2020) 'Framing "digital well-being" as a social good', *First Monday*. https://doi.org/10.5210/fm.v25i12.10430

Bengtsson, S. and Johansson, S. (2022) 'The meanings of social media use in everyday life: filling empty slots, everyday transformations, and mood management', *Social Media + Society*, 8(4). https://doi.org/10.1177/20563051221130292

Bericat, E. (2016) 'The sociology of emotions: four decades of progress', *Current Sociology*, 64(3), pp 491–513.

Boczkowski, P. J. (2021) *Abundance: On the Experience of Living in a World of Information Plenty*. Oxford University Press.

Bossio, D. and Holton, A. E. (2021) 'Burning out and turning off: journalists' disconnection strategies on social media', *Journalism*, 22(10), pp 2475–2492.

Bozan, V. and Treré, E. (2024) 'When digital inequalities meet digital disconnection: studying the material conditions of disconnection in rural Turkey', *Convergence*, 30(3), pp 1134–1148.

Brennen, B. (2019) *Opting Out of Digital Media*. Routledge.

Brubaker, R. (2023) *Hyperconnectivity and Its Discontents*. Wiley.

Bucher, T. (2020) 'Nothing to disconnect from? Being singular plural in an age of machine learning', *Media, Culture & Society*, 42(4), pp 610–617.

Caldeira, S. P. and Machado, A. F. (2023) 'The red lipstick movement: exploring #vermelhoembelem and feminist hashtag movements in the context of the rise of far-right populism in Portugal', *Feminist Media Studies*, 23(8), pp 4252–4268.

Chia, A. and Beattie, A. (2021) 'Ethics and experimentation in the light phone and Google digital wellbeing', in Chia, A., Jorge, A., and Karppi, T. (eds) Reckoning with Social Media: *Disconnection in the Age of the Techlash*. Rowman & Littlefield, pp 127–146.

Chia, A., Jorge, A. and Karppi, T. (2021) *Reckoning with Social Media: Disconnection in the Age of the Techlash*. Rowman & Littlefield.

Coleman, R. (2025) Infrastructures of Feeling: Digital Mediation, Captivation, Ambivalence. Working paper, Freie Universität Berlin. https://doi.org/10.17169/refubium-45893

Coleman, R. and Paasonen, S. (2020) 'Introduction: mediating presents', *Media Theory*, 4(2). https://doi.org/10.70064/mt.v4i2.636

Couldry, N. (2012) *Media, Society, World: Social Theory and Digital Media Practice*. Polity Press.

Couldry, N. and Hepp, A. (2017) *The Mediated Construction of Reality*, John Wiley & Sons.

Dahlgren, P. and Hill, A. (2020) 'Parameters of media engagement', *Media Theory*, 4(1). https://doi.org/10.70064/mt.v4i1.618

DataReportal (2025) *Digital 2025: Portugal*. https://datareportal.com/reports/digital-2025-portugal

Ehn, B., Löfgren, O. and Wilk, R. (2016) *Exploring Everyday Life: Strategies for Ethnography and Cultural Analysis*. Rowman & Littlefield.

Fast, K. and Syvertsen, T. (2024) 'Post-digital consumption: the controversy surrounding the mobile phone box as a means of disconnection', in Albris, K., Fast, K., Karlsen, F., Kaun, A., Lomborg, S. and Syvertsen, T. (eds) *The Digital Backlash and the Paradoxes of Disconnection*, Nordicom, University of Gothenburg, pp 45–66. https://urn.kb.se/resolve?urn=urn:nbn:se:norden:org:diva-13243

franzke, aline shakti, Bechmann, A., Zimmer, M., Ess, C. M. and Association of Internet Researchers (2020) *Internet Research: Ethical Guidelines 3.0*. Association of Internet Researchers. https://aoir.org/reports/ethics3.pdf

Frykman, J. and Povrzanovic Frykman, M. (2016) 'Affect and material culture: perspectives and strategies', in Frykman, J. and Povrzanovic Frykman, M. (eds) *Sensitive Objects: Affect and Material Culture*, Nordic Academic Press, pp 9–28. https://doi.org/10.21525/kriterium.6

Gibbs, A. (2011) 'Affect theory and audience', in Nightingale, V. (ed) *The Handbook of Media Audiences*. Blackwell Publishing, pp 251–66.

Gomes, A. and Dias, J. G. (2025) 'Digital divide in the European Union: a typology of EU citizens', *Social Indicators Research*, 176(1): 149–172.

Griffero, T. (2021) 'Are atmospheres shared feelings?', in Trigg, D. (ed) *Atmospheres and Shared Emotions*, Routledge, pp 17–39. https://doi.org/10.4324/9781003131298-2

Hermes, J. (1995) *Reading Women's Magazines: An Analysis of Everyday Media Use*. Polity Press.

Hermes, J. (2024) 'Introduction – Affect and identity: a cultural studies perspective on audience research', in Hill, A. and Lunt, P. (eds) *The Routledge Companion to Media Audiences*, Routledge, pp 349–359.

Highmore, B. (2002) *Everyday Life and Cultural Theory: An Introduction*. Routledge.

Highmore, B. (2011) *Ordinary Lives: Studies in the Everyday*. Routledge.

Hillis, K., Paasonen, S. and Petit, M. (eds) (2015) *Networked Affect*. MIT Press.

Illouz, E. (2008) *Saving the Modern Soul: Therapy, Emotions, and the Culture of Self-Help*. University of California Press.

Jansson, A. and Adams, P.C. (eds) (2021) *Disentangling: The Geographies of Digital Disconnection*. Oxford University Press.

John, N. and Katz, D. (2023) 'A history of features for online tie breaking, 1997–2021', *Internet Histories*, 7(3), pp 237–253.

Jolly, J. (2025) 'Spain and Portugal power outage: what caused it, and was there a cyber-attack?', *The Guardian*, 29 April. Available at: https://www.theguardian.com/business/2025/apr/28/spain-and-portugal-power-outage-cause-cyber-attack-electricity

Jorge, A. (2019) 'Social media, interrupted: users recounting temporary disconnection on Instagram', *Social Media + Society*, 5(4). https://doi.org/10.1177/2056305119881691

Jorge, A. (2023) 'Pilgrimage to Fátima and Santiago after COVID: Dis/connection in the post-digital age', *Mobile Media & Communication*, 11(3), pp 549–565.

Jorge, A. (2024) 'Dis/connected atmospheres: tourist locations in dead zones in post-pandemic Portugal', in Albris, K., Fast, K., Karlsen, F., Kaun, A., Lomborg, S. and Syvertsen, T. (eds) *The Digital Backlash and Paradoxes of Disconnection*. Nordicom, pp 325–344. https://doi.org/10.48335/9789188855961-16

Jorge, A., Amaral, I. and de Matos Alves, A. (2022) '"Time well spent": the ideology of dis/connection as a means for digital wellbeing', *International Journal of Communication*, 16, pp 1551–1572.

Jorge, A., Agai, M., Dias, P. and Martinho, L. C.-V. (2024) 'Growing out of overconnection: the process of dis/connecting among Norwegian and Portuguese teenagers', *New Media & Society*, 26(11), pp 6779–6795.

Karppi, T. (2018) *Disconnect: Facebook's Affective Bonds*. University of Minnesota Press.

Karppi, T., Chia, A. and Jorge, A. (2021) 'In the mood for disconnection', *Convergence*, 27(6), pp 1599–1614.

Kaun, A. and Treré, E. (2020) 'Repression, resistance and lifestyle: charting (dis)connection and activism in times of accelerated capitalism', *Social Movement Studies*, 19(5–6), pp 697–715.

Kennedy, A.K. (2018) 'The affective turn in feminist media studies for the twenty-first century', in Harp, D., Loke, J. and Bachmann, I. (eds) *Feminist Approaches to Media Theory and Research*. Palgrave Macmillan, pp 65–81.

Koivunen, A., Nikunen, K., Hokkanen, J., Jaaksi, V., Lehtinen, V., Soronen, A., Talvitie-Lamberg, K. and Valtonen, S. (2024) 'Anticipation as platform power: the temporal structuring of digital everyday life', *Television & New Media*, 25(2), pp 115–132.

Kolehmainen, M. and Mäkinen, K. (2021) 'Affective labour of creating atmospheres', *European Journal of Cultural Studies*, 24(2), pp 448–463.

Kuntsman, A. and Miyake, E. (2019) 'The paradox and continuum of digital disengagement: denaturalising digital sociality and technological connectivity', *Media, Culture & Society*, 41(6), pp 901–913.

Lefebvre, H. (1991) *Critique of Everyday Life*. Verso.

Lefebvre, H. (2004) *Rhythmanalysis: Space, Time and Everyday Life*. Continuum.

Lehto, M. and Paasonen, S. (2021) '"I feel the irritation and frustration all over the body": affective ambiguities in networked parenting culture', *International Journal of Cultural Studies*, 24(5), pp 811–826.

Light, B. (2014) *Disconnecting with Social Networking Sites*. Palgrave Macmillan UK.

Light, B. and Cassidy, E. (2014) 'Strategies for the suspension and prevention of connection: rendering disconnection as socioeconomic lubricant with Facebook', *New Media & Society*, 16(7), pp 1169–1184.

Lomborg, S. (2020) 'Disconnection is futile: theorizing resistance and human flourishing in an age of datafication', *European Journal of Communication*, 35(3), pp 301–305.

Lomborg, S. and Ytre-Arne, B. (2021) 'Advancing digital disconnection research: introduction to the special issue', *Convergence*. https://doi.org/10.1177/13548565211057518

Lunt, P. (2024) 'The felt experience of atmosphere: implications for audience research', in Hill, A. (ed) *The Routledge Companion to Media Audiences*, Routledge, pp 471–482.

Lupinacci, L. (2021) '"Absentmindedly scrolling through nothing": liveness and compulsory continuous connectedness in social media', *Media, Culture & Society*, 43(2), pp 273–290.

Lupinacci, L. (2024) 'Phenomenal algorhythms: the sensorial orchestration of "real-time" in the social media manifold', *New Media & Society*, 26(7), pp 4078–4098.

Lupinacci, L. (2025) 'Mixed feelings: the platformisation of moods and vibes', *AoIR Selected Papers of Internet Research*. https://doi.org/10.5210/spir.v2024i0.13992

Lupton, D. (2017) 'How does health feel? Towards research on the affective atmospheres of digital health', *Digital Health*, 3. https://doi.org/10.1177/2055207617701276

Mannell, K. (2019) 'A typology of mobile messaging's disconnective affordances', *Mobile Media & Communication*, 7(1), pp 76–93.

Markham, T. (2020) *Digital Life*. Wiley.

Maslen, S. (2022) 'Affective forces of connection and disconnection on Facebook: a study of Australian parents beyond toddlerhood', *Information, Communication & Society*, 26(9), pp 1716–1732.

Moe, H. and Madsen, O. J. (2021) 'Understanding digital disconnection beyond media studies', *Convergence*, 27(6), pp 1584–1598.

Mollen, A. and Dhaenens, F. (2018) 'Audiences' coping practices with intrusive interfaces: researching audiences in algorithmic, datafied, platform societies'. In Das, R. and Ytre-Arne, B. (eds) *The Future of Audiences: A Foresight Analysis of Interfaces and Engagement*. Springer, pp 43–60.

Mutsvairo, B., Ragnedda, M. and Mabvundwi, K. (2023) '"Our old pastor thinks the mobile phone is a source of evil": capturing contested and conflicting insights on digital wellbeing and digital detoxing in an age of rapid mobile connectivity', *Media International Australia*, 189(1), pp 89–103.

Nau, C., Zhang, J., Quan-Haase, A. and Mendes, K. (2023) 'Vernacular practices in digital feminist activism on Twitter: deconstructing affect and emotion in the #MeToo movement', *Feminist Media Studies*, 23(5), pp 2046–2062.

Paasonen, S. (2015) 'As networks fail: affect, technology, and the notion of the user', *Television & New Media*, 16(8), pp 701–716.

Paasonen, S. (2016) 'Fickle focus: distraction, affect and the production of value in social media', *First Monday*, 21(10). https://doi.org/10.5210/fm.v21i10.6949

Paasonen, S. (2021) *Dependent, Distracted, Bored: Affective Formations in Networked Media*. MIT Press.

Papacharissi, Z. (2014) *Affective Publics: Sentiment, Technology, and Politics*. Oxford University Press.

Petit, M. (2015) 'Digital disaffect: teaching through screens', in Hillis, K., Paasonen, S., and Petit, M. (eds) *Networked Affect*. MIT Press, pp 169–183.

Pink, S. and Leder Mackley, K. (2013) 'Saturated and situated: expanding the meaning of media in the routines of everyday life', *Media, Culture & Society*, 35(6), pp 677–691.

Pink, S., Sumartojo, S., Lupton, D. and Heyes La Bond, C. (2017) 'Mundane data: The routines, contingencies and accomplishments of digital living', *Big Data & Society*, 4(1). https://doi.org/10.1177/2053951717700924

Plaut, E. R. (2015) 'Technologies of avoidance: the swear jar and the cell phone', *First Monday*. https://doi.org/10.5210/fm.v20i11.6295

Portwood-Stacer, L. (2013) 'Media refusal and conspicuous non-consumption: the performative and political dimensions of Facebook abstention', *New Media & Society*, 15(7), pp 1041–1057.

Pype, K. (2021) '(Not) in sync – digital time and forms of (dis-)connecting: ethnographic notes from Kinshasa (DR Congo)', *Media, Culture & Society*, 43(7), pp 1197–1212.

Ribak, R. and Rosenthal, M. (2015) 'Smartphone resistance as media ambivalence', *First Monday*, 20(11). https://doi.org/10.5210/fm.v20i11.6307

Riedel, F. (2019) 'Atmosphere'. In Slaby, J. and von Scheve, C. (eds) *Affective Societies: Key Concepts*. Routledge, pp 85–95.

Rosenberg, H. (2019) 'The "flashpacker" and the "unplugger": cell phone (dis)connection and the backpacking experience', *Mobile Media & Communication*, 7(1), pp 111–130.

Salazar, M.M. (2023) *Cosy vibes: cosiness as an atmospheric aesthetic category*. Doctoral thesis (PhD), University of Sussex.

Sampson, T., Maddison, S. and Ellis, D. (2018) *Affect and Social Media: Emotion, Mediation, Anxiety and Contagion*. Rowman & Littlefield.

Silverstone, R. (1994) *Television and Everyday Life*. Routledge.

Stewart, K. (2011) 'Atmospheric attunements', *Environment and Planning D: Society and Space*, 29(3), pp 445–453.

Sumartojo, S. and Pink, S. (2018) *Atmospheres and the Experiential World: Theory and Methods*. Routledge. https://doi.org/10.4324/9781315281254.

Syvertsen, T. (2017) *Media Resistance – Protest, Dislike, Abstention*. Palgrave Macmillan.

Syvertsen, T. (2020) *Digital Detox: The Politics of Disconnecting*. Emerald Group Publishing.

Syvertsen, T. and Enli, G. (2020) 'Digital detox: Media resistance and the promise of authenticity', *Convergence: The International Journal of Research into New Media Technologies*, 26(5–6), pp 1269–1283.

Syvertsen, T. and Ytre-Arne, B. (2021) Privacy, energy, time and moments stolen: social media experiences pushing towards disconnection. In Chia, A., Jorge A. and Karppi, T. (eds) *Reckoning with Social Media*. Roman & Littlefield, pp 85–102.

Thorén, C., Edenius, M., Lundström, J. E. and Kitzmann, A. (2019) 'The hipster's dilemma: what is analogue or digital in the post-digital society?', *Convergence*, 25(2), pp 324–339.

Tucker, I. M. and Goodings, L. (2017) 'Digital atmospheres: affective practices of care in Elefriends', *Sociology of Health & Illness*, 39(4), pp 629–642.

Wahl-Jorgensen, K. (2019) 'Intimacy, emotion, and journalism', in Wahl-Jorgensen, K. (ed) *Oxford Research Encyclopedia of Communication*, Oxford University Press. https://doi.org/10.1093/acrefore/9780190228613.013.823

Weltevrede, E., Helmond, A. and Gerlitz, C. (2014) 'The politics of real-time: a device perspective on social media platforms and search engines', *Theory, Culture & Society*, 31(6), pp 125–150.

Wetherell, M. (2012) *Affect and Emotion: A New Social Science Understanding*. Sage.

Williams, R. (1977) *Marxism and Literature*. Oxford University Press.

Ytre-Arne, B. (2023) *Media Use in Digital: Everyday Life*. Emerald Publishing Limited.

Ytre-Arne, B., Syvertsen, T., Moe, H. and Karlsen, F. (2020) 'Temporal ambivalences in smartphone use: conflicting flows, conflicting responsibilities', *New Media & Society*, 22(9), pp 1715–1732.

ONE

Post-Digital Parenting: The Relational-Affective Network of the Family

Francisca Porfírio, Ana Jorge and Rita Grácio

Introduction

Today, portable media are not confined to the home, and the home is also permeated by permanent connectivity and emergent technologies (Leaver, 2015; Kennedy et al, 2020). Children and families are situated in an increasingly mediatized – platformized, datafied and algorithmic – world (Mascheroni and Siibak, 2021; Das, 2024; Sefton-Green et al, 2025), and parents must tackle children's inevitable 'digital future' (Livingstone and Blum-Ross, 2020).

The role of media in everyday life and the home has long been the object of reflection and inquiry in different disciplines (Pink and Leder Mackley, 2013; Chambers, 2016; Ytre-Arne, 2023). However, most research on families and media has focused on how parents perform the 'good parent' through parental mediation, notably drawing on domestication theory (Clark, 2014). There has been growing attention to how the media participate in constructing the family identity. Digital parenting (Mascheroni et al, 2018) has been conceptualized as encompassing both parental mediation and practices by parents that include digital media in daily life. Parenting-related activities on digital media include looking for information online or what Lim calls 'transcendent parenting' (2020), micro-managing children's activities; as well as sharenting, the representation of parenting on social media (Ong et al, 2022; Esfandiari and Yao, 2023).

The family has been conceptualized as an intimate and relational network (Roseneil and Ketokivi, 2016) that is constantly being built through everyday, shared practices and activities (Thimm, 2023). The socioconstructivist approach that understands the family as a product of its interactions, relationships, forms of organization and daily practices is the basis of the definition of 'doing family' (Rose, 2016; Scheibling, 2020; Zerle-Elsäßer, 2023). However, besides studies focusing on parents (Cino and Vandini, 2020; Lehto and Paasonen, 2021; Das, 2024), there is still scarce attention to the role of media considering the 'emotionally-laden relational interactions' (Clark, 2014, p 32) and affective aspects of the families (Pink and Leder Mackley, 2013).

This chapter primes the role of emotions and moods in the ever-changing network of family relations surrounding digital media. To this end, we turn to conceptualizations of 'post-digital'. Fast proposed that post-digital parenting includes the labour that parents perform in securing the 'digital health of family members' (2021, p 1623), amid the normalization of transcendent parenting or digital parenting. To Fast, this labour, which is instrumental to the platform economy, is increasingly incorporated into parenting norms, though it is highly gendered and classed, and notably includes tasks related to screen-time control. In an earlier conceptualization of the term, Cramer associates the post-digital with the latent 'sentiments of disenchantment and scepticism' towards digital technology (2015, p 19). These sentiments, which can be more vague or more concrete, can explain an attitude of 'participatory reluctance' (Cassidy, 2018), whereby individuals resort to digital media cynically. Post-digital can also be understood as a synthesis between the digital and non-digital (Taffel, 2016);

or hybrid artefacts that blur digital and analogue (Thorén et al, 2019). In previous work (Jorge, 2023), we found digital practices to which individuals ascribed non-digital meanings (for example, sending an email intended as a postcard) as evidence of the post-digital era.

We propose here to extend the notion of post-digital parenting to include the emotional and affective elements in articulating the mediation of children's engagement with (rather than use of) digital media in the home, and the online representation and datafication of family. We emphasize how these processes are riddled with tensions, disagreements and continuous negotiations through a network of affective relations, interweaving the findings from our two-part qualitative study. Our analysis focuses on how media are part of 'doing family', the affective labour involved in negotiations surrounding digital media, and relational and digital labour that extends family in relations with others. Our data from ethnography and content analysis revealed the nuances, oscillations and hesitations in negotiations that involve underlying values and norms of how the family is being continuously 'done', but also build embodied and situated agreements of what media represent in everyday home atmospheres, and beyond.

Materials and methods

This chapter draws on a two-part study on family life and digital media: a content analysis of eight parents' accounts on Instagram and an ethnography with five families in Portugal. Our Instagram study included eight public accounts by Portuguese parents, characterized in Table 1.1, found through parenting-related hashtags. The selected cases were four mothers and four fathers who did not have an influencer status. Instagram is the focal social media platform as it is popular among Portuguese adults (Statista, 2025) and widely thought of as a 'photo album for children' (Choi and Lewallen, 2018, p 1). We considered content that mentioned or alluded to children or parenting in the three formats on Instagram, namely posts, stories and reels, during 2022; given the amount of Instastories, we considered only content from 100 days of the year, selected randomly with PineTool. We selected all posts and reels published in 2022 for analysis. Our final corpus was 437 posts, 168 reels and 508 Instastories. Content was downloaded with Insta-Save, or screenshot when this option was unavailable, organized and saved on a hard drive only accessible to the authors. From disclosed occupations and cues regarding lifestyle, the selected parents' accounts suggest that they have middle socioeconomic profiles.

For the ethnographic study, we sought five Portuguese families with at least one child up to 12 years old, from diverse situations in terms of education and occupation, family forms and living arrangements, geography, and migrant status (for example, Harden et al, 2010; Gouveia and Castrén, 2021): families Almeida, Barroso, Carvalho, Dias and Esteves. We collected data through a survey questionnaire, observation on two family life occasions, and interviews with parents and children aged 8 or older. In the spring of 2023, we recruited participants with an e-flyer advertised on the researchers' social media accounts, and each participating family was compensated with a voucher. Observation occurred on two occasions, three hours each, on regular days and during a family meal (Chen et al, 2019), with only notes being produced, and not audio or video. On a third visit, interviews were conducted with parents (as a couple/pair, except for the single parent in family Barroso), and with children 8 or older; as elicitation, interviews used a photo of the participants' favourite device(s) which was requested at the end of the second visit. A total of around nine hours of interviews were transcribed verbatim.

We recruited 20 participants, including nine adults and seven children 8 years old or above who were interviewed. The families were composed of two parents except in the Barroso family, where Ana was a single mother after a divorce; there was one child in the Carvalho family, two children in

Table 1.1: Parents on Instagram – sample

Pseudonym	Number of followers	Occupation	Hashtag/search term	Children and age (years)	Notes
Filipa	562	Stay-at-home mum	#maesdeportugal (mothersofportugal)	1 boy (2)	Announced new pregnancy in February 2022
Gabriela	3,818	Owner of small business	#momlife	2 boys (5 and 8)	–
Helena	2,491	Flight attendant	#lovemyfamily	1 girl (1)	–
Inês	662	Nurse	#babydiary	1 boy (1)	Daughter to be born in February 2022
Joaquim	1,086	Photographer and digital manager	#myson	1 boy (2)	–
Leonardo	592	Kindergarten teacher	#primeirofilho (firstson)	1 boy (8 months)	Announced his partner's new pregnancy in December 2022
Manuel	1,319	Unknown	Search for 'pai' (father)	1 boy (3)	Announced his partner's new pregnancy in May 2022
Nuno	10.9 m	Unknown	#daddy	1 boy (1)	–

Notes: The number of followers was collected in February 2021, when selecting accounts for a larger PhD study; number and age of children in January 2022, at the start of data collection for this study.

families Almeida, Barroso and Esteves, and four children in family Dias – the youngest from their current common-law marriage, two from the mother's and one from the father's previous relationship. In this family, Júlio, the father, was an immigrant, whereas the rest of the participants were born in Portugal. The Almeida and Barroso families were from a rural part of central mainland Portugal, whereas the other three lived in the urban area of Greater Lisbon. Given their qualifications and occupations (see Rose and Harrinson, 2007), our participants had a lower socioeconomic status, except for Carvalho family, who had the highest education and skilled occupations, placing them as middle socioeconomic status family. This profile may have resulted from the voucher compensation offered for participation, considered low value for upper-middle socioeconomic status families. Nevertheless, we believe this granted us access to a social group of parents that is not often researched and can complement analyses of privileged (Andelsman, 2024) or low-income families (Mannell et al, 2024).

The project's codebook was adapted to each of the parts of the study, through inductive and deductive reasoning (DeCuir-Gunby et al, 2011) and with an iterative discussion among the authors and the project's team. Analysis was conducted on qualitative software MaxQDA, with a multimodal approach for the Instagram material (Bouvier and Rasmussen, 2022) and as thematic analysis for the ethnographic data (Clarke and Braun, 2017).

For the Instagram material, we considered publicly available content to be subjectable to analysis, and consent was not obtained for observation; nevertheless, we followed the Association of Internet Researchers' (AoIR) *Ethical Guidelines* (franzke et al, 2020): we adopted first name pseudonyms for the accounts and anonymized the children, and discarded other identifiable elements from the data. For the ethnographic study, a consent form, with information about the study, and the voluntary, anonymous and confidential nature of the data collected, as well as the right to withdraw, was signed by parents and was orally adapted for children; we chose pseudonyms (popular Portuguese surnames for ABCDE) to reference each of the five participant families (Gurrieri and Drenten, 2019), and individual pseudonyms were chosen by the participants or the authors.

Doing family in the digital media age

Through our ethnography, we found families inserted in highly mediatized environments (Zerle-Elsäßer, 2023), who use media as part of 'doing family' in different ways. Young Clara Almeida, aged 9, expresses how she is like every other member of the family spending time online: 'There are games on Roblox. My mom [installed] this, my brother [installed] it and my dad [installed] it and I have one too and so we all play.' Media also help to connect with extended family and cement family ties, as father João Esteves indicates: 'With the family, we usually always talk on camera. Even to see the kids, so it's part of it ... It's also a way of bonding, seeing each other.' Social media also allows families to show their moments to others, be it family, friends or strangers, as Graça Carvalho tells us: 'If it weren't for this way [social media], some of the people I know wouldn't know Diana.'

As found in other studies (Koniski, 2018), despite the unequivocal presence of media, and particularly digital media, in families' everyday lives (Chambers, 2016; 2021), offline family leisure activities are presented by most parents in both of our studies as a preferable way for children to spend their free time. For our participants, these activities included 'going out for walks, painting, ... cleaning as a family', as Ana Barroso indicates; or 'playing board games', according to Lina Esteves. Lina's partner, João, also expresses that parents should make an effort to 'introduce them [children] to other interests' besides technology. João's words indicate a moral economy where digital technology is what is more easily available and should be countered. These positions from our participants of middle socioeconomic status resonate with popular discourses about what makes 'good parenting' as often oriented to improving the health and quality of life of the family (Knibb and Taylor, 2017; Mäkinen, 2021), which can be seen as more highbrow (Torras-Gómez et al, 2021).

The same cultural hierarchy was present when parents referred to media activities, showing their attempt to promote positive media effects, including by preferring educational content and preventing negative content (Lemish, 2015; Livingstone and Blum-Ross, 2020; Mascheroni and Siibak, 2021). Our interviewed parents valued and recognized the benefits of consuming digital content, mainly through learning applications recommended by schools, such as the Hypatiamat platform, mentioned by the Almeida family, or Google Classroom, mentioned by the Esteves family. Such cultural hierarchies can also help explain the scarcity of mentions of media on Instagram discourses, where only around 4 per cent of our corpus was related to media, mirroring social norms associated with 'good parenting' as disconnected. When media appear in their discourses on Instagram, parents often show themselves interacting with the platform and their followers. Only in rare instances do they show their children interacting with digital devices – but when this happens, it is mostly in playful settings. Digital media are portrayed in use in public spaces, such as restaurants, as shown in one Instagram picture in a carousel by Miguel (30 July 2022) where the child is sitting at a restaurant table with a tablet in front of him; this may help to maintain the child's good behaviour and prevent crying or tantrums that could disturb others; through heart-eyes emojis in the caption ('😍😍😍 [smiling face with heart-eyes]'), the father frames this as an affectionate moment, rather than a behaviour that is often socially sanctioned.

Indeed, in contrast with normative positions on digital health that discourage children's use of media during meals and for long periods, we found instances where digital media act as an important tool to regulate children's emotions, or a *pacifier*. Ana Barroso illustrates how media are of help when either the parent or the child is tired: 'Sometimes it happens and I don't know what to do, so I grab the mobile phone, give it to her and she can be on the sofa for hours … So, it saves my mental health. She's quiet, so it helps the harmony of the house a little.'

Children we interviewed reported enjoying offline activities, such as going on family walks, playing ball, or going to the beach. However, they also enjoy using digital media (for example, playing online games), particularly at home. These media activities are often performed with other family members. Nine-year-old Clara Almeida told us: 'I like to play PS [PlayStation] with my brother'; and 11-year-old Margarida Barroso recounted: 'We have a game on the Nintendo Switch, which is about cars and we [she and her mother] used to compete to see who could do the least amount of time', which indicates co-participatory digital mediation (Livingstone and Helsper, 2008).

In the ethnographic sample, all but one parent stated that they shared content online about their children, mainly on Facebook and Instagram, confirming the normalization of sharenting. The exception was João Carvalho, 41, who has no social media. In all five families, the primary motivations for parents to share about their children on social media were displaying pride in their children's progress and receiving feedback from others. Most of all, parents expressed a liking to share offline family activities, often in outdoor contexts, 'We post when we go somewhere, the four of us. To the Estádio da Luz [Portuguese team Benfica's football stadium], for example', Hugo Almeida, 44, told us. Among our sample, parent–children conflict or negative feelings over parental sharing (Blum-Ross and Livingstone, 2017) were not reported, unlike previous studies which found that this could be the case if children perceived that their image was used embarrassingly or incorrectly (Ouvrein and Verswijvel, 2019). Speaking in a third-person scenario to express his position, 8-year-old André Esteves says:

> If it is going to be published and his son feels confident, and if his son says, 'I feel confident about you publishing it', yes. But if he doesn't and it's embarrassing, I think the person, the child, would be a little sad with their parents if they [the child] found out about it.

This example may indicate that parents have incorporated recommendations about seeking consent from their children before sharenting. Clara, aged 9, testifies that children and pre-adolescents approve and feel recognized when their parents publish about their achievements; to the girl, this is augmented by the fact that 'there is no one in the world who can't see these photos'.

Through our multimodal analysis, Instagram appeared as a stage for showcasing family leisure and harmony, primarily through offline family interaction activities – similar to those described by our interviewees, such as photos on a day out. This family interaction is mainly present in routine contexts, but also in commemorative dates or cultural, recreational and sporting events and activities. Another frequent representation across our corpus was the moments spent between parents and children without technology. For example, in one Instagram reel by Helena (16 November 2022), the child is sitting on her mother's lap while painting; the caption signals a frame of bonding: 'Mother-daughter moments ♥ [red heart] … #daughter #mother #moments #motherdaughter #painting #love' (translated from Portuguese). Again, there is an investment in displaying parent–child interactions around non-mediated activities, even if produced by a smartphone and shared via social media. Another of these non-mediated activities that were often represented in content produced by parents was mealtimes, both routine and on special occasions, appearing as a way to take a break from the media to strengthen family ties.

The family investment around mealtime, as dissociated from the media, was also present in four of the (five) families participating in our ethnographic study. As 16-year-old Daniel (Dias) points out: 'When I'm sat at the table …, none of us have our phone, we're eating, chatting about everyday life, things like that'. From this perspective, the meal appears among these Portuguese families as a moment to take a break from digital media and bring family members closer together. In the Esteves family, the mother, Lina, tells us that 'during normal school hours, you know, after doing his homework, André can go to his technology and, at dinner time, it's over. After dinner, there is no more technology for either of us.' This seemed to be especially the case on school evenings.

The exception to dissociating meals from media was the Barroso family, where the single mother, Ana, told us that they always watched the *Once Upon a Time* series on Disney+ together while having dinner because she believes that watching a television series with her daughters can help to strengthen family ties between them. Ana and her two girls said they would often go on TikTok for tips on activities with diamond painting, which uses tiny, diamond-like sparkles to create colourful designs and patterns. They exemplify a hybrid activity inspired by online content to be conducted offline, which can be seen as a post-digital instance. Furthermore, the ambivalence of media is also evident in the fact that, as we will see next, Ana also controls her children's use of digital media out of fear of exposing them to risks.

Tensions, disagreements and affective labour

Ana Barroso, 32 years old and a single mother, considers herself a cautious mother and is an avid user of the parental control app Family Link to monitor her daughters' digital activity. She told us that once, when checking her 11-year-old daughter's WhatsApp, she identified that Margarida was in a group chat with a stranger, remembering that she had a 'terrible experience as a mother, because you feel like you can't control everything ... They are at a difficult age, in the sense that they don't even rebel against us but think that they are the ones who are right'. She also reported using a family chat group for 'day-to-day things, for example, when she comes home from school, I always like her to tell me where she is'. Many parents who use apps to monitor or track their children, or their parental care activities (Stoilova et al, 2023; Sefton-Green et al, 2025), report feeling safer. However, these apps pose concerns regarding continuous surveillance as 'an enhanced parenting tool' (Marx and Steeves, 2010, p 205) and the very data policies of parental control platforms are not transparent to parents (Barassi, 2020). When asked, Ana Barroso says she 'cannot control' how third parties use data. However, this mother seems to trade the absence of information on data management off for the immediate sense of control and safety.

Her daughter, Margarida, recalled in our interview how her mother was once 'upset' with her for spending too much time playing computer games. Children's engagement with media challenges parents, particularly in terms of usage, frequency of use and what they perceive to be the future implications for children (Clark, 2011; Eichen et al, 2021). This can lead to interactions that can become conflictual (Koniski, 2018) and even take the form of arguments and fighting (Kennedy et al, 2020, p 180). In the Dias family, it is 12-year-old Guilherme who reveals that he has experienced some tense moments between him and his stepmother, Dulce, remembering that she sometimes gets upset with him for spending too much time on his phone: 'She gets mad for, I don't know, a day. And then forgets about it.' Guilherme also illustrates how children navigate the rules regarding digital technologies constantly but also flexibly, that is, depending on different environments of norms and rules (Plowman, 2016). At his mother's house, the time to use his phone is non-negotiable and he obeys her more than he does his stepmother: 'My mother says turn it off and I turn it off.' In the interview with Júlio and Dulce, both mentioned that managing technology in a home with four children is complex; Dulce told us that sometimes the last resort is: 'I get there and basically take the children's phones away from them [laughs].' Drastic measures like this enforce disconnection on the children, deriving from a necessity to keep the home running, as Mannell and colleagues (2024) found among Australian low-income families. When asked how the children react when this happens, Júlio indicates that they do not always accept it well; they get upset and sometimes even cry, especially the younger ones, Ângelo (5 years old) and Lisa (11). This example demonstrates that not everyone is equal in the shared atmosphere, that the interactions can quickly change that same atmosphere (Griffero, 2021), and that expressions such as sulking or crying might be the means by which children try to affect the atmosphere.

In such situations, the conflicts between children's and parents' desires reinforce hierarchical family power relations and moral responsibility (Koniski, 2018). But this does not stop children, according to Lina Esteves, mother of two boys, from continuously trying to use the media if they so desire. Furthermore, Lina also reveals that children's interactions are always adapted to whom they interact with – and that other family members also transform the environments. She says: 'They always try to stretch the limit to where they can go and if it's with me, it's one thing ... If it's with their father, it's different. If grandma is here, they know they can get around it another way.' Here, the 'limit' refers not just to screen time but to any rule that adults have for children, and that she perceives the children acknowledge, but push back on. The father, João, also told us that access to digital technologies by children needs to be 'filtered', but, as the mother sees it, children interact with each parent differently around media use.

Interactions between siblings around media also reveal affective ductility, as they can be both supportive and conflictual. This is demonstrated by the relationship between young Simão Almeida, 16, and his younger sister, Clara, 9.

On the one hand, he says he scolds her for posting videos on TikTok, while leaving issues of concern and safety to their parents. On the other hand, he says his younger sister bothers him as she keeps wanting to see what he is doing on his phone or computer.

Different from the Esteves couple, we observed greater dissonance among the couple João and Graça Carvalho, parents of a baby girl, not even one year old at the time of our study, especially around the topic of sharenting. Like other parent couples who disagree among themselves (Clark, 2014, p 323), Graça likes to share moments with her baby, Diana, on her social media accounts, while João, during the couple's interview, explained his position as a compromise: 'If it were up to me, Diana wouldn't exist on the internet. However, Diana is part of our lives, part of Graça's life. And Graça wants to share part of her life. It's done with common sense and in agreement.'

Despite adopting different positions regarding the practice of sharenting among themselves, they agreed to warn Diana's paternal and maternal grandparents not to expose their granddaughter's image. Grandparents play a considerable role in their grandchildren's lives and frequently use social media (Staes et al, 2023), and here again conflicts may arise. Graça and João pre-emptively announced their decision, which holds that only the baby's mother is allowed to share about her on social media. Parents seem to reclaim 'first-level agency in managing their children's digital footprints' (Cino and Vandini, 2020, p 181), over the extended family.

On the flip side, some tensions may be worked through and social support found through social media (Lehto and Paasonen, 2021; Mackenzie, 2024). Instagram appears as a digital space where parents, especially mothers (Archer and Kao, 2018), vent about their feelings of guilt, internal dilemmas and social expectations about parenting (Porfírio and Jorge, 2022; Scheibling and Milkie, 2023; Tartari et al, 2023). Our Instagram study found instances where parents reflect on their struggle to meet complex parental demands while highlighting the parent–child connection. This is illustrated by a reel (13 September 2022) where Filipa, a stay-at-home mother, mirrors through in-video text a dilemma associated with parenting, to express that engaging with children and family time comes before any other task, although maintaining and organizing the house seems to be her pending duty. In the caption, the mother reinforces this idea: 'They're the best thing in my life, at the end of the day they're all that matters, the house and everything else are just things, and in a few years we won't even remember that, but they'll always be here and they'll remember that we were there 🥹 [face holding back tears].' This type of affective negotiation is in line with the conclusions of a study of 20 mothers about the publication of family photographs on social media, who did not display idealized images of the family and discourses of 'good' motherhood but instead attempted to repair problematic identities in highly complex ways through the selection of the photographs they posted (Lazard, 2022).

As much as it allows for emotional labour, sharenting entails its own complex mix of challenges and risks, notably when parents are scrutinized and judged for their parenting choices (Thimm, 2023). For example, Graça Carvalho confided:

> Once I posted an instastory and a girl sent me a message asking if I'd ever thought about correcting Diana's ears ... It's someone who knows me on the internet, but doesn't have a relationship with me, like ... I went to the doctor ... I know you're upset with me (looking at her partner), but it's not intended to be mean, you know.

In this situation, Instagram became a source of judgement towards her daughter, which Graça felt was inappropriate because the person did not know her personally. Social media opens up parents to unwanted scrutiny that fuels feelings of insecurity and hurt, and requires affective labour (Mäkinen, 2021; Lehto and Paasonen, 2021; Lim and Wang, 2024). In Graça's case, the discomfort was aggravated because the father does not sharent. Graça also described another situation where she was offered unsolicited medical advice after sharing about the child being ill, when the parents had already sought treatment for her. Furthermore, Graça's experiences indicate that, even when parents choose to mobilize positive, rather than negative, affect in their sharenting (Lazard, 2022), audience reactions can be unexpected and disturbing.

Relational and digital labour

Besides managing children's digital presence and positioning their conduct in the universe of care and safety (Andelsman, 2024), parents maintain connections with children through multiple apps and platforms, while cultivating other networked relationships around parenting. As we explore, all these forms of connection require a relational labour that demands skills (Beuckels et al, 2024) that ultimately reconfigure the meanings of the interactions with the family (Scribano and Lisdero, 2019). Furthermore, this labour is not formally recognized, and it especially affects mothers in our sample, a pattern that can be found in more privileged contexts (Andelsman, 2024).

Among our sample, we could see that 'transcendent parenting' (Lim, 2016; 2020) is motivated by the incorporation of the ideology of intensive parenting, which reflects a scenario where children's needs and desires tend to be prioritized over those of their caregivers (Faircloth, 2023). Being continually aware of children's lives, whether through monitoring apps and parental control features (Ghosh et al, 2018), such as Family Link, or even through immediate control actions, such as checking children's mobile phones after use, gives parents feelings of security and 'peace of mind' (Sandra Almeida). For example, mothers Ana Barroso and Dulce Dias reported regularly checking their children's smartphones to ensure that they are digitally safe. Dulce says, 'I know it might be a little bit of an invasion of their privacy, but it's because of his age', and Ana reflects how 'It's very hard to find a balance, right?

Her [Margarida's] privacy with her security, because sometimes I have to ..., sometimes I have to go over her privacy a little to protect her.'

Intimate surveillance, a concept described by Leaver (2015) to refer to digital surveillance practices adopted by well-intentioned child-carers, is visible from a very early age. The parents of infant girl Diana, Graça and João Carvalho, pointed out that the kindergarten offers a 'very useful' app, where the staff logs data on her physiological needs during the day (feeding, sleeping, bowel issues, and so on). Over our visits, the parents told us that this helped them understand 'certain reactions from her' (Notes, 27 June 2023), and also manage product stocks; for example, they would know when they need to bring nappies to the kindergarten. Although quantifying children's basic and physiological needs practices is not new, the 'appification' of these practices represents significant changes in how childcare is carried out and understood (Sefton-Green et al, 2025). By embracing the app offered by the school, João and Graça, the most qualified among our ethnography's participants, reveal their eagerness to participate in the child's personal and school life and to strive to enhance the child's well-being that defines the norm of 'good parent' (Cino et al, 2021), while trying to facilitate routines. At the same time, the app supports them in managing the family's emotional well-being as they care for an infant.

Connecting with peer parents appeared as an important motivation for sharenting among parents on Instagram. In our Instagram analysis, besides the moments of joy when children are happy and reflect family unity that we described earlier, ordinary parents also shared challenges and frustrations associated with sleep, the difficulty of routines, nutrition and breastfeeding. We could detect that parents, especially mothers, on Instagram often resort to relatability as an affective mechanism to invite the audience to identify with them, using storytelling to build trust (Atiq et al, 2022). Inês used this form of digital labour: she talked on her Instagram account about her frustration and difficulty in the first two weeks of her daughter's life because she was unable to breastfeed, and turned breastfeeding into a frequent topic in her posts. Besides this, like other parents, she uses a lighter tone and funny captions, using emojis, which help to create a sense of closeness and community and relieve the pressure of perfect parenting. One example of this was a post (19 August 2022) with the caption 'snacks by the river 👨 👩 👧 👦 [family: man, woman, girl, boy]' of a selfie taken by the husband, where she is breastfeeding her daughter outdoors on a sunny summer day, and the older son is between the couple; the adults are looking at the camera and smiling. On another post (18 December 2022), she posted a photo of herself and another mother with their babies; in the caption, she indicates that this is the first time they have met in person after talking online for a long time, 'staying up until 1 am' because 'the girls are crazy and won't sleep'. She mentions that they supported each other during pregnancy, postpartum, breastfeeding difficulties and other challenges.

Among our ethnography participants, contact with other parents and social groups related to children's activities through instant messaging was mostly mentioned by mothers, and associated with overwhelm and pressure always to be reachable, if seen in combination with the availability for children's contacts. Dulce Esteves describes how 'we have so many [chat] groups, in my case, I have so many groups … sometimes we're out for a walk and I'm answering one mother here and then I'm calling another'. Mothers perform relational labour on behalf of their children: continuous labour, without schedules, which depends on online interaction, and extends women's 'second shift' (Hochschild and Machung, 1989). The other face of this pressure is that, when they are away from their children and technology fails for some reason, mothers feel anxious and restless, as Andelsman (2024) also found among Danish parents and Jorge (2024) among Portuguese parents and carers in touristic dead zones. For example, Ana Barroso tells us 'When I'm not with them [children], if, for some reason, the battery runs out or there's no signal … that makes me a bit anxious … I'll only feel calm when my phone is working properly again.'

Conclusion

Adopting a post-digital framework, this chapter offered a comprehensive perspective on the affective relations between parents and children, as well as the wider family, about and around digital media. Our combination of qualitative studies looking at different samples offers insights into families of middle socioeconomic status in a South European context, where 'good parenting' discourses seem to be enacted in both family online representations and norms of mediation of use, which reserve a place for digital media that is ambivalent for parents. These moral economies are superseded and must be adapted by situational decisions, for example trying to prevent fights or avoid bothering someone in a public space with a child's tantrum.

Our analysis revealed that joy and happiness, as much as stress and anxiety, can arise from parents' and children's encounters with digital media (Lehto and Paasonen, 2021). Positive emotions are often overlooked by research in children and media, and we believe that seeing those in combination with what can be perceived as negative emotions is crucial to understanding the larger picture of the affective interactions that take place in the home – and the extended family. Our material showed how social media allows parents to work on emotions and expose dilemmas, find support and create meaningful connections, as well as opening the possibility for disturbance from people outside the home, or disagreement between the parents, for which compromises might be made. Digital media afford parents feelings of assurance that their children are safe, or grant them 'peace of mind', but they largely overlook privacy concerns in sharenting and data shared on platforms. Engaging in transcendent parenting depends on digital and relational labour performed mainly by mothers, who can easily get overwhelmed. This continuous labour makes parents feel they are not allowed to disconnect, as they have particular pressure to be available through phone, instant messaging and so on, for matters concerning their children. While connected to a larger family network, the parents and the nuclear family reclaim the leading agency to oversee children's use of digital media and their digital footprint, and pre-emptively draw boundaries.

Furthermore, digital media may pacify or entertain children, help parents anticipate children's behaviour, or make the family feel together, while they can also be at the centre of chaos in the home, cause conflict or motivate mood changes. Decisions about media use are not just emotional and relational (Clark, 2014; Das, 2024), but also subject to more prescient sensations or dispositions such as being tired after a workday, feeling bothered by a younger sibling looking over one's shoulder, or respecting a mother more than a stepmother. Children overusing digital media may be preferred over having them cry and throw tantrums, but mediating such overuse might also induce crying. Nevertheless, our qualitative-based chapter can only offer a glimpse into families in south-western Europe, and therefore we hope to see more research considering how digital media and families interact in everyday home atmospheres.

References

Andelsman, V. (2024) 'Navigating the moral imperatives of parenting in the age of (dis)connection: a care-minded approach to digital media use by parents in Denmark', in Albris, K., Fast, K., Karlsen, F., Kaun, A., Lomborg, S. and Syvertsen, T. (eds) *The Digital Backlash and Paradoxes of Disconnection*. Nordicom, University of Gothenburg, pp 235–256. https://doi.org/10.48335/9789188855961-12

Archer, C. and Kao, K.-T. (2018) 'Mother, baby and Facebook makes three: does social media provide social support for new mothers?' *Media International Australia*, 168(1), pp 122–139.

Atiq, M., Abid, G., Anwar, A. and Ijaz, M. F. (2022) 'Influencer marketing on Instagram: a sequential mediation model of storytelling content and audience engagement via relatability and trust', *Information*, 13(7), p 345.

Barassi, V. (2020) *Child Data Citizen*. MIT Press. https://mitpress.mit.edu/9780262044714/child-data-citizen/

Beuckels, E., Hudders, L., Vanwesenbeeck, I. and Van den Abeele, E. (2024) 'Work it baby! A survey study to investigate the role of underaged children and privacy management strategies within parent influencer content', *New Media & Society*, 27(6), pp 3081–3101. https://doi.org/10.1177/14614448231218992

Blum-Ross, A. and Livingstone, S. (2017) '"Sharenting," parent blogging, and the boundaries of the digital self', *Popular Communication*, 15(2), pp 110–125.

Bouvier, G. and Rasmussen, J. (2022) *Qualitative Research Using Social Media*. Routledge.

Cassidy, E. (2018) *Gay Men, Identity and Social Media: A Culture of Participatory Reluctance*. Routledge.

Chambers, D. (2016) *Changing Media, Homes and Households: Cultures, Technologies and Meanings*. Routledge.

Chambers, D. (2021) 'Emerging temporalities in the multiscreen home', *Media, Culture & Society*, 43(7), pp 1180–1196.

Chen, Y.-Y., Li, Z., Rosner, D., and Hiniker, A. (2019) 'Understanding parents' perspectives on mealtime technology', *Proceedings of the ACM on Interactive, Mobile, Wearable and Ubiquitous Technologies*, 3(1), pp 1–19.

Choi, G. Y. and Lewallen, J. (2018) '"Say Instagram, kids!": Examining sharenting and children's digital representations on Instagram', *Howard Journal of Communications*, 29(2), pp 144–164.

Cino, D. and Vandini, C. D. (2020) '"My kid, my rule": Governing children's digital footprints as a source of dialectical tensions between mothers and daughters-in-law', *Studies in Communication Sciences*, 20(2), pp 181–202.

Cino, D., Gigli, A. and Demozzi, S. (2021) '"That's the only place where you can get this information today!": an exploratory study on Parenting WhatsApp Groups with a sample of Italian parents', *Studi sulla Formazione/Open Journal of Education*, 24(1), pp 75–96.

Clark, L. S. (2011) 'Parental mediation theory for the digital age', *Communication Theory*, 21(4), pp 323–343.

Clark, L. S. (2014) 'Mobile media in the emotional and moral economies of the household', in Goggin, G. and Hjorth, L. (eds) *The Routledge Companion to Mobile Media*. Routledge, pp 320–332.

Clarke, V. and Braun, V. (2016) 'Thematic analysis', *The Journal of Positive Psychology*, 12(3), pp 297–298.

Cramer, F. (2015) 'What is "post-digital"?', in Berry, D. M. and Dieter, M. (eds) *Postdigital Aesthetics: Art, Computation and Design*. Palgrave, pp 12–26.

Das, R. (2024) *Parents Talking Algorithms: Navigating Datafication and Family Life in Digital Societies*. Bristol University Press.

DeCuir-Gunby, J. T., Marshall, P. L. and McCulloch, A. W. (2011) 'Developing and using a codebook for the analysis of interview data: an example from a professional development research project', *Field Methods*, 23(2), pp 136–155.

Eichen, L., Hackl-Wimmer, S., Eglmaier, M. T. W., Lackner, H. K., Paechter, M., Rettenbacher, K., Rominger, C. and Walter-Laager, C. (2021) 'Families' digital media use: intentions, rules and activities', *British Journal of Educational Technology*, 52(6), pp 2162–2177.

Esfandiari, M. and Yao, J. (2023) 'Sharenting as a double-edged sword: evidence from Iran', *Information, Communication & Society*, 26(15), pp 2942–2960.

Faircloth, C. (2023) 'Intensive parenting and the expansion of parenting', in Lee, E. Bristow, J., Faircloth, C. and Macvarish, J. (eds) *Parenting Culture Studies*. Springer International Publishing, pp 33–67.

Fast, K. (2021) 'The disconnection turn: three facets of disconnective work in post-digital capitalism', *Convergence*, 27(6), pp 1615–1630.

franzke, aline shakti, Bechmann, A., Zimmer, M., Ess, C.M. and Association of Internet Researchers (2020) *Internet Research: Ethical Guidelines 3.0.* Association of Internet Researchers. https://aoir.org/reports/ethics3.pdf

Ghosh, A. K., Badillo-Urquiola, K., Guha, S., LaViola Jr, J. J. and Wisniewski, P. J. (2018) 'Safety vs. surveillance: what children have to say about mobile apps for parental control', in *Proceedings of the 2018 CHI Conference on Human Factors in Computing Systems.* Montreal QC, Canada, 19 April 2018. ACM. doi.org/10.1145/3173574.3173698

Gouveia, R. and Castrén, A.-M. (2021) 'Redefining the boundaries of family and personal relationships', in Castrén, A.-M., Česnuitytė, V., Crespi, I., Gauthier, J.-A., Gouveia, R., Martin, C., Moreno, A. M. and Suwada, K. (eds) *The Palgrave Handbook of Family Sociology in Europe.* Springer International Publishing, pp 259–277. https://doi.org/10.1007/978-3-030-73306-3_13

Griffero, T. (2021) 'Are atmospheres shared feelings?', in Trigg, D. (ed) *Atmospheres and Shared Emotions*, Routledge, pp 17–39. https://doi.org/doi.org/10.4324/9781003131298-2

Gurrieri, L. and Drenten, J. (2019) 'Visual storytelling and vulnerable health care consumers: normalising practices and social support through Instagram', *Journal of Services Marketing*, 33(6), pp 702–720.

Harden, J., Backett-Milburn, K., Hill, M. and MacLean, A. (2010) 'Oh, what a tangled web we weave: experiences of doing "multiple perspectives" research in families', *International Journal of Social Research Methodology*, 13(5), pp 441–452.

Hochschild, A. and Machung, A. (1989) *Working Parents and the Revolution at Home.* Viking.

Jorge, A. (2023) 'Pilgrimage to Fátima and Santiago after COVID: dis/connection in the post-digital age', *Mobile Media & Communication*, 11(3), pp 549–565.

Jorge, A. (2024) 'Dis/connected atmospheres: tourist locations in dead zones in post-pandemic Portugal', in Albris, K., Fast, K., Karlsen, F., Kaun, A., Lomborg, S. and Syvertsen, T. (eds) *The Digital Backlash and Paradoxes of Disconnection.* Nordicom, pp 325–344.

Kennedy, J., Arnold, M., Gibbs, M., Nansen, B. and Wilken, R. (2020) *Digital Domesticity: Media, Materiality, and Home Life*. Oxford University Press.

Knibb, J. N. and Taylor, K. (2017) 'Living "light green": the limits and lived experiences of green motherhood', *Qualitative Market Research: An International Journal*, 20(3), pp 370–389.

Koniski, E. (2018) '"Please turn it off": negotiations and morality around children's media use at home', *Discourse & Society*, 29(2), pp 142–159.

Lazard, L. (2022) 'Digital mothering: sharenting, family selfies and online affective-discursive practices', *Feminism & Psychology*, 32(4), pp 540–558.

Leaver, T. (2015) 'Born digital? Presence, privacy, and intimate surveillance', in Hartley, J. and Qu, W. (eds) *Re-Orientation: Translingual Transcultural Transmedia. Studies in Narrative, Language, Identity, and Knowledge*. Fudan University Press, pp 149–160.

Lehto, M. and Paasonen, S. (2021) '"I feel the irritation and frustration all over the body": affective ambiguities in networked parenting culture', *International Journal of Cultural Studies*, 24(5), pp 811–826.

Lemish, D. (2015) *Children and Media: A global perspective*. Wiley.

Lim, S. S. (2016) 'Through the tablet glass: transcendent parenting in an era of mobile media and cloud computing', *Journal of Children and Media*, 10(1), pp 21–29.

Lim, S. S. (2020) *Transcendent Parenting: Raising Children in the Digital Age*. Oxford University Press. https://doi.org/10.1093/oso/9780190088989.001.0001

Lim, S. S. and Wang, Y. (2024) 'Social media and performative parenting', in Skoric, M. M. and Pang, N. (eds) *Research Handbook on Social Media and Society*. Edward Elgar Publishing, pp 2–11. https://doi.org/10.4337/9781800377059.00009

Livingstone, S. and Helsper, E .J. (2008) 'Parental mediation of children's internet use', *Journal of Broadcasting & Electronic Media*, 52(4), pp 581–599.

Livingstone, S. and Blum-Ross, A. (2020) *Parenting for a Digital Future: How Hopes and Fears about Technology Shape Children's Lives*. Oxford University Press.

Mackenzie, J. (2024) '(Dis)connected parenting: context control and information management in single adoptive parents' social media practice', *Adoption & Fostering*, 48(2), pp 203–222.

Mäkinen, K. (2021) 'Resilience and vulnerability: emotional and affective labour in mom blogging', *New Media & Society*, 23(10), pp 2964–2978.

Mannell, K., Boyle, E., Kennedy, J. and Holcombe-James, I. (2024) '"Taking the router shopping": how low-income families experience, negotiate, and enact digital dis/connections', *New Media & Society*. https://doi.org/10.1177/14614448241234941

Marx, G. and Steeves, V. (2010) 'From the beginning: children as subjects and agents of surveillance', *Surveillance & Society*, 7(3/4), pp 192–230.

Mascheroni, G. and Siibak, A. (2021) *Datafied Childhoods: Data Practices and Imaginaries in Children's Lives*. Peter Lang.

Mascheroni, G., Ponte, C. and Jorge, A. (eds) (2018) *Digital Parenting: The Challenges for Families in the Digital Age*. Nordicom.

Ong, L. L., Fox, A. K., Cook, L. A., Bessant, C., Gan, P., Hoy, M. G., Nottingham, E., Pereira, B. and Steinberg, S. B. (2022) 'Sharenting in an evolving digital world: increasing online connection and consumer vulnerability', *Journal of Consumer Affairs*, 56(3), pp 1106–1126.

Ouvrein, G. and Verswijvel, K. (2019) 'Sharenting: Parental adoration or public humiliation? A focus group study on adolescents' experiences with sharenting against the background of their own impression management', *Children and Youth Services Review*, 99, pp 319–327.

Pink, S. and Leder Mackley, K. (2013) 'Saturated and situated: expanding the meaning of media in the routines of everyday life', *Media, Culture & Society*, 35(6), pp 677–691.

Plowman, L. (2016) 'Rethinking context: digital technologies and children's everyday lives', *Children's Geographies*, 14(2), pp 190–202.

Porfírio, F. and Jorge, A. (2022) 'Sharenting of Portuguese male and female celebrities on Instagram', *Journalism and Media*, 3(3), pp 521–537.

Rose, G. (2016) *Doing Family Photography: The Domestic, the Public and the Politics of Sentiment*. Routledge.

Rose, D. and Harrison, E. (2007) 'The European Socio-economic Classification: a new social class schema for comparative European research', *European Societies*, 9(3), pp 459–490.

Roseneil, S. and Ketokivi, K. (2016) 'Relational persons and relational processes: developing the notion of relationality for the sociology of personal life', *Sociology*, 50(1), pp 143–159.

Scheibling, C. (2020) 'Doing fatherhood online: men's parental identities, experiences, and ideologies on social media', *Symbolic Interaction*, 43(3), pp 472–492.

Scheibling, C. and Milkie, M. A. (2023) 'Shifting toward intensive *parenting* culture? A comparative analysis of top mommy blogs and dad blogs', *Family Relations*, 72(2), pp 495–514.

Scribano, A. and Lisdero, P. (2019) 'Work and sensibilities: commodification and processes of expropriation around digital labour', in Scribano, A. and Lisdero, P. (eds) *Digital Labour, Society and the Politics of Sensibilities*. Springer International Publishing, pp 39–60. https://doi.org/10.1007/978-3-030-12306-2_3

Sefton-Green, J., Mannell, K. and Erstad, O. (eds) (2025) *The Platformization of the Family: Towards a Research Agenda*. Springer Nature Switzerland. https://doi.org/10.1007/978-3-031-74881-3

Staes, L., Walrave, M. and Hallam, L. (2023) 'Grandsharenting: how grandparents in Belgium negotiate the sharing of personal information related to their grandchildren and engage in privacy management strategies on Facebook', *Journal of Children and Media*, 17(2), pp 192–218.

Statista (2025) Portugal: social media users by age group and platform 2024. Available at: https://www.statista.com/statistics/1373430/portugal-social-media-users-by-age-group-and-platform/

Stoilova, M., Bulger, M. and Livingstone, S. (2023) 'Do parental control tools fulfil family expectations for child protection? A rapid evidence review of the contexts and outcomes of use', *Journal of Children and Media*, 18(1), pp 29–49.

Taffel, S. (2016) 'Perspectives on the postdigital: beyond rhetorics of progress and novelty', *Convergence*, 22(3), pp 324–338.

Tartari, M., Lavorgna, A. and Ugwudike, P. (2023) 'Share with care: negotiating children's health and safety in sharenting practices', *Media, Culture & Society*, 45(7), pp 1453–1470.

Thimm, C. (2023) 'Mediatized families: digital parenting on social media', in Dethloff, N., Kaesling, K., and Specht-Riemenschneider, L. (eds) *Families and New Media: Comparative Perspectives on Digital Transformations in Law and Society*. Springer, pp 33–57.

Thorén, C., Edenius, M., Lundström, J. E. and Kitzmann, A. (2019) 'The hipster's dilemma: what is analogue or digital in the postdigital society?', *Convergence*, 25(2), pp 324–339.

Torras-Gómez, E., Ruiz-Eugenio, L., Sordé-Martí, T. and Duque, E. (2021) 'Challenging Bourdieu's theory: dialogic interaction as a means to provide access to highbrow culture for all', *Sage Open*, 11(2). https://doi.org/10.1177/21582440211010739

Ytre-Arne, B. (2023) *Media Use in Digital: Everyday Life*. Emerald Publishing Limited.

Zerle-Elsäßer, C. N., Langmeyer, A., Naab, T. and Heuberger, S. (2023) 'Doing family in the digital age', in Skopek, J. (ed) *Research Handbook on Digital Sociology*. Edward Elgar Publishing, pp 365–378. https://doi.org/10.4337/9781789906769.00030

TWO

Platformized Feminisms and Social Media Ambiences

Sofia P. Caldeira, Ana Jorge and Ana Kubrusly

Introduction

As digital and social media platforms have become embedded in everyday life, contemporary feminist and activist practices have increasingly been imagined as 'digital by default' (Fotopoulou, 2016). Social media has helped to expand activisms beyond the purview of traditional actors (like activists, collectives or organizations), inviting the participation of 'ordinary' individuals (Milan, 2015, p 59). It has facilitated the creation of feminist communities (Kanai and McGrane, 2021) and the expansion of feminist knowledge cultures (Kanai, 2021). This turn towards the use of digital technologies for feminist purposes – for information sharing, advocacy, movement-building, organization and mobilization, among others (Özkula, 2021) – has been framed as a shift towards a fourth wave of feminism (Munro, 2013). In this context, being politically active on social media becomes enmeshed with one's sense of feminist identity and fosters feelings of connectedness with others, simultaneously prompting fears of being left out or forgotten if not engaging in these practices (Fotopoulou, 2016).

These feminist potentialities coexist with critiques of potential depoliticization of activisms on digital media, which risks detaching the over-abundant communication of, and about, politics from concrete and politically efficient actions (Dean, 2005). Digital activist practices can sometimes be derided as *slacktivism* (Özkula, 2021; Ponder et al, 2023), perceiving online activism as more passive, less labour-intensive and less impactful than traditional forms of activism.

These criticisms can draw on the fallacy of 'digital dualism', understanding the digital as separate from (and, at times, lesser than) the material realm of 'real' life (Özkula, 2021). However, as our everyday experiences repeatedly attest, the digital and the non-digital cannot be taken as entirely separate realms, as online experiences can spill over and affect our offline lives and moods – and vice versa. These dynamics can also be felt in digital activism, complicating its definition (Özkula, 2021). This chapter explores the relationships between feminisms and social media ambiences, recognizing that these movements do not entirely revolve around media (Mattoni, 2017, 2020) and focusing on how people live with and affectively experience social media in feminist contexts.

Engaging with digital feminist activisms can often be an intense affective experience (Mendes, 2022), both in contexts that disrupt the ordinariness of everyday life – coordinated movements such as #MeToo, or strikes, marches, or street protests – and in *everyday feminisms* (Pruchniewska, 2019), within the same social media platforms (for example, Facebook, Instagram) that are already widely used for other purposes and 'naturalized' in quotidian life (Campos et al, 2018). Thus, practices of *platformized feminism* (Barbala, 2024) become entangled, shaped by and negotiated with each platform's features, rules, conventions of use and communities of practice.

Using the framework of affective atmospheres (see Introduction) and exploring the impacts of platformized feminisms as a paradigmatic example, we propose the notion of *ambience* to narrow down on the specificity of social media platforms – the practices that communities build around each specific platform, as well as what in those experiences may call for disengagement. *Ambience*, explored in geography and tourism as well as cultural studies, not only foregrounds ideas of space and refers to the background of experiences that may be imperceptible, but also refers to how people attune to these spaces, that is, how mediated spaces affect people's perceptions, moods, actions and social experiences (Roquet, 2012; Lim et al, 2020). As ambiences can play a role in shaping political experiences (Khilnani, 2021), this also invites an exploration of the ambivalent and fleeting relations established with platforms, thus allowing us to engage with both the issues and the potential of these platforms for activist practices (van Zoonen, 2011).

Our proposed notion of ambiences seeks to make sense of everyday experiences with social media as encounters between each user's moods and feelings, and platforms, as a composite of vernacular cultures, affordances and their technological, sociocultural and economic contexts. Experiences and encounters with social media are affected not only by the wider cultural environment (Paasonen, 2021), but also by 'infra-structures of feeling' (Coleman, 2018), which can be reflected in pre-emergent affective states that compel or lure us to engage with these platforms. The architectures of different platforms create different possibilities for interaction (Papacharissi, 2009), but ambiences are generated between this transmission of affect and the individual experience that can change it (Tucker and Goodings, 2017).

As this chapter explores, relationships with social media platforms are also affected by less tangible elements, such as platform *vibes*. Popularized in digital vernacular discourse, vibes rely on 'common sensations' that are affectively mediated, highlighting the perceptions and moods collectively associated with certain platforms (Lupinacci, 2025). These platform vibes are not stable; rather, they are maintained through digital labour of both users and platforms themselves, and can be strategically incorporated and even capitalized by social media platforms (Kolehmainen and Mäkinen, 2021; Brown et al, 2024; Lupinacci, 2025).

Context and study: 2023 International Women's Day march

This chapter explores these everyday activist dynamics in Portugal – a context where the empirically based scholarship on contemporary and digital activisms is still sparse (Campos et al, 2018, p 494). After the end of dictatorship in 1974, Portugal's feminist modern history has been dominated by practices of 'state feminism', rather than by grassroots movements and bottom-up mobilizations (Santos and Pieri, 2019). In this context, although platforms like Facebook (Marôpo et al, 2017; Campos et al, 2018) or, more recently, Instagram (Lamartine and Cerqueira, 2023) have been used for activist practices, digital feminist movements and international hashtag campaigns, such as the #MeToo movement, have struggled to gain a strong public expression nationally (Garraio et al, 2020) or local efforts were met with low metrics of engagement (Marôpo et al, 2017). In recent years, however, there has been a growth in on- and offline feminist mobilizations in response to political shifts and controversies (for example, Caldeira and Machado, 2023; Cerqueira et al, 2023). Among these, we include the yearly International Women's Day marches (Lamartine and Cerqueira, 2023).

Taking place yearly on 8 March, the International Women's Day march – or the 8M march – began in Portugal in 2017, gaining momentum and public attention in 2019, under the leadership of the March 8 network. Since then, it is considered one of the most expressive feminist mobilizations within the Portuguese context, notably using Instagram to organize and disseminate initiatives (Lamartine and Cerqueira, 2023). The collectives who organize the march and the broader movement strategically use their online presence to increase their relevance (Fernández-Romero and Sánchez-Duarte, 2019; Lamartine and Cerqueira, 2023), while simultaneously highlighting the symbolic importance of occupying the streets (Gago, 2020). We use the 8M march in 2023 as a methodological entry point for our study.

Considering that on- and offline feminist engagements are deeply entangled, we adopt a multi-methodological and multi-situated approach. In line with the overall project's strategy, we combine three strategies for data collection: online observations on Instagram of the organization and representation of the 8M march; offline ethnographic observations of the 8M march in Lisbon; and in-depth, semi-structured interviews with people engaged with the feminist movement in the Portuguese context, exploring participants' experiences both off- and online and their affective impact.

On Instagram, observations were conducted in the two weeks leading up to the 2023 8M march, on the day of the protest, and on the following day. We used PhantomBuster to automatically extract data regarding three Instagram profiles of feminist collectives involved in 8M mobilization in Portugal, and the hashtag #FeminismoPortugal. This strategy generated a corpus of 637 posts and 354 stories.

Four researchers documented the protest in Lisbon with audiovisual recordings and textual fieldnotes. As informed consent could not be secured for all protest participants, all materials collected from online and offline observations were visually anonymized and used for internal analysis only.

The recruitment strategy for interview participants was multi-situated. During the march, researchers distributed flyers inviting protesters to participate. Moreover, we invited (via private message on Instagram) some of the people who created or engaged with the materials collected in our online observations. Additionally, as a snowball technique, we asked interviewees to recommend other potential participants, whom we invited. In the end, we conducted 22 interviews between March and May 2023. These included the participation of people with sustained activist engagements as well as 'ordinary' people (in other words, not activists, career politicians, or celebrities) who engage with feminist and political topics more sporadically. Table 2.1 provides an overview of participants. Interviews

Table 2.1: Overview of the interviewees

Name/pseudonym*	Age	Gender	Nationality	Occupation	Political activities	Recruited from
Ana Fidalgo	32	F	Portuguese	Academic researcher	Organizes feminist conversation circles and maintains an Instagram page to promote them	Instagram post from hashtags
Ana Sofia	44	F	Portuguese	Geographer / Cultural mediator	Affiliated to a political party	Instagram post from hashtag
Ana Vaz	34	F	Portuguese	Financial risk analyst	Affiliated to a political party and member of political collectives	8M march
Bruno Leitão Alves	35	M	Portuguese	Restaurant owner	Founder of a communist Discord group that organizes online mobilizations	Instagram stories
Camila	34	F	Brazilian	Academic researcher	Member of political collectives; studies feminist movements	Instagram stories
Carlota	27	F	Portuguese	Sign language teacher and content creator	Everyday feminist **	Instagram stories

Table 2.1: Overview of the interviewees (continued)

Name/ pseudonym*	Age	Gender	Nationality	Occupation	Political activities	Recruited from
Carolina Fernandes	23	F	Portuguese	Student	Member of political collectives; founder of an anti-harassment movement, promoted through an Instagram page	Instagram stories
Catarina	26	F	Portuguese	Graphic designer, content creator	Creates political digital content	Instagram stories
Catarina Silva	26	F	Portuguese	Journalist	Everyday feminist	Instagram stories
Danielle Tavares	50	F	Brazilian	Psychologist, content creator	Creates political digital content	Instagram post from hashtag
Ester de Oliveira	22	F	Portuguese	Student	Member of an anti-harassment movement, promoted through an Instagram page	Instagram stories
Lena Hertel	20	F	Portuguese and German	University student	Member of political collectives	Instagram stories
Marisa Carboni	61	F	Italian and Brazilian	Costume designer	Affiliated to a political party and member of political collectives	8M march

Table 2.1: Overview of the interviewees (continued)

Name/ pseudonym*	Age	Gender	Nationality	Occupation	Political activities	Recruited from
Marta	29	F	Portuguese	Communication specialist	Feminist activist (not affiliated to any collective)	Instagram stories
Marta D	36	F	Portuguese	Communication freelancer	Member of political collectives. Creates political digital content	Snowball
Miguel Martins	22	M	Portuguese	Student	Municipal deputy. Member of political collectives. Organizes political events	Instagram stories
Mourana Monteiro	26	F	Portuguese	Community intervention technician / university student	Member of political collectives	Instagram stories
Pedro Miguel Santos	24	M	Portuguese	Graphic designer	Affiliated to a political party. Creates digital political content for the party	Instagram stories
Rita Oliveira	26	F	Portuguese	Project manager	Creates political digital content	Instagram stories

Table 2.1: Overview of the interviewees (continued)

Name/pseudonym*	Age	Gender	Nationality	Occupation	Political activities	Recruited from
Sofia Pinto	39	F	Portuguese	Biomedical engineer	Organizes feminist conversation circles and maintains an Instagram page to promote them	Instagram post from hashtags
Telmo Pinto	34	M	Portuguese	Photojournalist	Everyday feminist	Instagram stories
Yasmin	27	F	Malaysian	Freelancer	Everyday feminist	Instagram stories

Notes: * These refer to the interviewee's name or, if they chose to remain anonymous, their self-chosen pseudonyms; gender: F = female, M = male as self-identified by the participants. ** Everyday feminist refers to those participants who engage with feminist politics in their daily lives, without affiliation to any collective or organization.

were all conducted online, at the participants' convenience, averaging 50 minutes.

After transcription, all data collected through the different techniques were aggregated using MaxQDA. Data were thematically coded (Legard et al, 2003): following an in-depth familiarization with the data, we followed an iterative and theoretically informed coding process, identifying emerging themes within the data and establishing relationships between them (Gaskell, 2006). The coding process was collectively discussed among the involved researchers.

Given the topic's sensitive nature, ethical precautions were also taken to inform all research participants about the research aims and potential risks, requiring their informed consent to participate. Participants could choose to participate anonymously in the research, under a pseudonym of their choosing, or to be identified by their first or full names. Only three participants chose to use a pseudonym. Regarding positionality, all researchers involved in this work are women, from Portugal and Brazil, albeit all working in a Portuguese context. We acknowledge that we share a feminist political alignment with the studied topic and research participants.

Digital by default, ambivalent at best

Although in-person organizing and protesting have not been entirely replaced by online habits (Fotopoulou, 2016), as the massive adherence to the 8 March marches showcases, several interviewees struggled to imagine contemporary feminisms *without* social media:

> I feel that, nowadays ... [hesitates] it's not like we can only have offline feminism if we have online feminism. The way to mobilize people and collectives goes through the online. So, I feel that, if we didn't have that, it would be much harder to reach people and talk with them. (Ester)

Social media platforms, like Instagram, were seen as one of the most important tools in the activist repertoire – allowing activists to reach new audiences, communicate, organize internally, make contact with other organizations, or mobilize large numbers of people for events. The use of these platforms is so naturalized and ingrained (Coleman, 2018) that, for Ester, it felt 'impossible' to imagine having the necessary reach without them.

This centrality of social media for feminist organizing was also reflected in our observations of Instagram in the weeks leading to the 2023 8M march, as the platform was heavily used to promote the event. Hashtags like #ACaminhoDo8M (on our way to 8M) were used in posts related to the preparation of the event and built expectation around the event (like what is presented in Chapter 3 for World Youth Day). Posts with Instagrammable graphic flyers and posters (Dumitrica and Hockin-Boyers, 2023; Caldeira, 2024) invited people to join the 'purple wave' of protesters across Portugal, with feminist messaging highlighting the importance of attending – 'together we are stronger'. These messages were often mixed with practical information like dates, times, meeting points in different Portuguese cities, or general recommendations for protesters.

Combining online and offline observations foregrounded the hybrid character of feminist actions like the 8M march. For instance, on Instagram, the material aspects of the march were highlighted, with photos and videos of workshops dedicated to crafting signs and banners being shared through Instagram stories. However, during the march itself, the digital remained ever present. Digital devices were widely used to document the protest and share through social media on the go. Digital objects (for example, QR codes, hashtags, social media username handles) were also present in material offline aspects of the protests, such as banners and flyers. Overall, practices building up to, after and during the march itself highlighted activists' hybrid practices based on flows between on- and offline activities, which mutually reinforce each other.

Participants highlighted additional advantages of using social media platforms, such as the ease of accidentally stumbling upon useful political content while scrolling (Ana Fidalgo) or how these platforms could facilitate a sense of feminist community. As Marisa put it, when questioned about the value of being present on social media: 'At least I don't stay completely isolated, right?', which seems to point at some reluctance in using the same platforms despite recognizing their value (Cassidy, 2018). For many participants, social media could provide *reassurance* and an awareness of belonging to a diffuse collective of people who shared similar experiences and feelings (Lupinacci, 2021, p 282), as Mourana reflected:

> It's like when you encounter a comrade in the middle of a crowd of people who are alienated from the issues, you know? That is, when I see people sharing in their stories, making posts, videos about [feminist] issues, it gives me hope. It gives me that feeling that I'm not alone, that there are more people who think like us. That feeling is very comforting and mobilizing.

These feelings of community can transcend social media into offline settings, as with Marta D., who joined the 8M network and their offline assemblies after finding their Instagram page. By engaging with wider feminist communities online, participants can not only find a sense of community but also feel reassured in their own feminist identities, invoking Markham's (2021) theory of social echolocation, which frames digital sociality as contributing to identity formation. This ability to encounter like-minded people on social media was seen as particularly important for those interviewees who lived outside of urban settings where offline feminist organizing is rarer, as Ana Sofia exemplified: 'We know how rural realities are like and that they are really different from a community point of view. When one lives in smaller towns [hesitates], one has more "exotic" communities, because the people who care about these things are very few.'

In this way, many participants positively experienced social media platforms, as they facilitated the creation of *feminist filter bubbles* (Kanai and McGrane, 2021) where one could more safely engage with people who share similar ideals and feed this collective energy. This comfort and safety can be facilitated by algorithms (Şot, 2022). Users rely on social media for this contagion (Sampson et al, 2018). At the same time, these bubbles could also be ambivalently perceived as narrowing down one's experience and isolating people from conflicting opinions, sometimes creating unrealistic expectations of the world. As Ana Sofia reflected: 'It's that feeling that we are all worried about the same things, right? So why aren't things working? And then we realize that there aren't that many of us, are there?'

The feelings of community facilitated by these platforms can thus be fluid and fleeting, relying on an affective experience of identification with others who share similar experiences, more than on sustaining enduring social connections (Lupinacci, 2021, p 282).

Indeed, people's experiences with and perceptions of these platforms for activist practices can be ambivalent and deeply situated, drawing on an interplay between people's affective states and social media ambiences. These experiences are, as was introduced earlier, not only shaped by and negotiated with the affordances and cultures of use of each platform (Gibbs et al, 2015; Barbala, 2024), but, as each platform gets imbued with particular symbolic dimensions and ever-changing political imaginaries (Mattoni, 2020), also dependent on more intangible platform *vibes* (Sampson et al, 2018; Kender, 2022; Brown et al, 2024; Lupinacci, 2025). Social media platforms are symbolic spaces of interaction, thus inherently social and dynamic. This idea has been earlier encapsulated as *textures* (Sørensen, 2009) and communicative *genres* (Lomborg, 2011) that platforms display and are characterized by.

The perceived characteristics and communities of each platform can thus push people towards (or against) specific platforms for their feminism, and can equally shape the kinds of practices carried out in them (Keller, 2019). Experiences of platformized feminism cannot be universalized, and their perceived advantages cannot be taken in isolation. Instead, as our interviews illustrate, these experiences can be polysemic, among the community and for each individual user.

Various participants recognized Twitter as a political platform par excellence – where one could find news, varied political opinions and lively discussions (Catarina S., Danielle, Rita). However, echoing commonly shared and intuitively felt perceptions of Twitter (Kender, 2022), nearly all participants saw this platform as having the most toxic and hateful ambience – due to its distinct audiences and cultures of use, short-text format, algorithmic influences, or governance, notably shifts following the acquisition by Elon Musk in 2022. As the example of Twitter showcases, both the material underpinning of these platforms and shifts in the symbolic meanings associated with them (Mattoni, 2020) can alter the affective investments and intensities they evoke.[1] This, as Keller (2019) notes, opens space for a fluidity of movements of feminist engagements across an ever-changing social media landscape. These platform ambiences caused ambivalent feelings in interviewees like Rita, who pondered: 'I think I could leave any platform, but not Twitter ... Twitter can be a toxic place, and I can fully recognize that there is a lot of malice there. But it's also, you see, a way to keep myself up to date with everything.'

Facebook, on the other hand, was generally understood as an outdated platform, at times useful for reaching an older audience that does not usually use other platforms (Catarina, Lena, Marisa). Facebook could thus be a point of contact with one's family and hometown, allowing interviewees like Lena to engage with collectives and associations from her rural hometown. Here, our interviewees resonate the idea that the ambience is created by all of those that take part in the interaction, who by their very personae as well as the content of their interactions affect the atmosphere – similar to the 'sophisticated ambiences' created by micro-celebrities' performances, as pointed out by Usher (2020).

Moreover, the connotation of its user community converged with the platform's algorithmic structure in creating a sense that Facebook was a more conservative space. Existing algorithmic imaginaries (Bucher, 2017) made several participants feel that Facebook favours anger to create more engagement (Merrill and Oremus, 2021), generating particular affective encounters that evoke different experiences and moods. Ana Sofia, Bruno, Carolina and Miguel felt there was an algorithmic push to confront them with content shared by people who might not share the same political ideals, thus making hateful interactions more likely. These experiences can complicate already ambivalent feelings towards the feminist filter bubbles encountered in digital contexts dominated by like-minded content.

> Facebook was the first platform where I felt that the algorithm was against me ... I started feeling that all the content I was shown was made by people with whom I argued. I wasn't shown things that interest[ed] me, photos from my friends, none of that ... It was the first platform where I felt: 'Ok, they are really using my anger for the profit of the platform.' (Bruno)

In turn, Instagram was seen as an ambivalent platform for activist practices. While for some, like Lena or Catarina, it was the platform where they connected with other activists and got news and information about upcoming events, for others, like Bruno, Instagram evoked associations with more personal and light-hearted content that could be seen as a waste of time. Unlike the affective discomfort caused by shared perceptions of Twitter or Facebook as toxic or annoying (Kender, 2022),

Instagram tended to be perceived as a *less toxic* environment (Duffy and Hund, 2019, p 4984). Although less pleasant experiences existed, Instagram tended to be experienced as a fun and aesthetically pleasing platform that generated, as Ana Sofia noted, a greater sense of 'ease' in its use, due to the less confrontational nature of its interactions. This 'peaceful bubble' offered by Instagram was in part facilitated by affordances like the one-to-one private communication enabled by Instagram stories. This exemplifies how platform affordances can be adopted, and adapted, to create a sense of privateness and community that is essential for constructive feminist discussion but can also be a self-preservation strategy, filtering out anti-feminist misogyny (Kanai and McCrane, 2021). Facilitating protective boundaries, these uses thus evoke strategies to foster safe(r) spaces that have long been part of feminist practices, as seen in historical semi-private spaces of offline feminist assemblies or consciousness-raising groups (Clark-Parsons, 2018).

Our online observations also illustrated how participants perceived and appropriated Instagram's affordances, shaping their activist engagements. For instance, tagging other accounts in stories can be a way of amplifying reach, allowing for others to re-share content. These functionalities can help both formal organizations and 'everyday' feminists to articulate themselves. This was exemplified by the peak of Instagram stories observed on the day after the 8M march, which included content created by 'everyday feminists' and re-shared by feminist collectives. Notably, this was also a way for collectives with local representations across Portugal to coordinate their offline efforts in the digital sphere, bringing coherence to the movement.

Similarly, our interviewees saw liking and commenting as ways of showing support: for example, Ana Fidalgo and Ana Vaz reported that they deliberately focused on engaging only with posts they agreed with to amplify the reach of positive messages. As Ana Sofia, Lena and other participants noted, re-sharing affordances could also serve political purposes (Caldeira, 2024), being used as implicit shows of support for causes or even to shape platforms' algorithmic structures. Pedro Miguel explains: 'On TikTok, the thing is to re-share it [feminist content], to make it more accessible.' Several multimodal functions used on Instagram stories – such as stickers, GIFs and songs – afforded an aesthetic valorization of photos and videos, emphasizing its feminist messaging. Locations, event reminders and countdowns leveraged the dissemination of the march across the country – even in smaller cities – and built up expectations surrounding the event. Unlike traditional media (Andiloro, 2023), on social media, the atmospheric assemblage is a set of possibilities that users appropriate, recreate, subvert – or altogether ignore.

Transient moods, pervasive pressures and responsibilities

The social media ambiences that surround these encounters are attributable not only to the platforms' social texture (Sørensen, 2009) and technological or institutional foundations, but also to users' different moods and attunements (Karppi et al, 2021). The ways in which interviewees engaged or disengaged with feminisms on these social media platforms were highly dependent on the moods and expectations users brought with them at any given moment. As Lena exemplified, her response towards feminist content depended largely on her current state of mind. When in a 'good' mental state, encountering news about an attack on a trans person or a case of gendered violence could feel galvanizing, motivating her into action and to start debates on the topic with people around her. In those moments, social media could be perceived as a valuable tool for political engagement. Yet, in the days she already feels despondent, the same negative content often serves to exacerbate her negative mood, leading her to feel like retreating and disengaging. In those days, as Catarina's description reasserts, these platforms seem to emphasize the dreadfulness of experience and thus can appear useless, or even toxic: 'There are days in which

I'm more at peace with the world and when, maybe, I'll see a post and I think: "Good content, I'll share it." Then, on other days when I'm more frustrated maybe I'll look at that post and think: "This world is really a disaster."'

These experiences are thus highly situated, transient and changing in intensity. Furthermore, what this shows is that, while social media ambiences are affected by pre-existing moods, these encounters can also, in turn, *affect* people's mood (Tucker and Goodings, 2017). Social media encounters can feed into different emotions, potentially resulting in ambivalent experiences (Lupinacci, 2021), particularly when dealing with highly affectively charged feminist and political topics. To Van Der Wal and colleagues (2024), such ambivalence is perceived as concurrent duality of emotions, while also alternate or sequential dualities can occur when young people engage with social media. In turn, Bengtsson and Johansson (2022, pp 8–9) note that social media use has been seen as facilitating mood management, as users reach for it in search of particular feelings during their everyday lives. Among respondents, these emotions were accumulated and not completely controllable. While social media could evoke positive feelings, it could also trigger undesirable feelings, for example when encountering high volumes of negative information: 'I spend a lot of time on social media. And, sometimes, I feel overwhelmed with all the information I'm receiving. And sometimes I even feel anguished. Because so much of the information we read is also bad' (Ester).

The ambiences of these platforms can also more directly affect how people engage or retreat from them, as well as their perceived political potential. As several interviewees mentioned, platforms like Twitter or Facebook can be experienced as toxic and particularly prone to online hate. There seems to be a growing *toxic technoculture* dominating social media (Massanari, 2017), which upholds toxic ideas and pushes against progressive social issues through implicit or explicit harassment. This can be reflected in an anti-feminism trend, which has accompanied the growth of popular feminist visibility (Banet-Weiser, 2018). As such, being visible as a feminist or activist on social media is not risk-free (Keller, 2019, p 9).

For interviewees, these practices seem to carry a looming threat of online hate – even if not necessarily experienced personally. For Miguel, 'hate comments are normal, unfortunately, and exhausting'. Nevertheless, as we expanded in earlier work (Caldeira et al, 2025), feelings of political responsibility can still compel people to contravene their own feelings of discomfort and obstinately remain in these toxic environments. Miguel continued, 'it's exhausting, of course, but that doesn't mean we should stop'. The ambiences of social media platforms can also play a role in this conflicting push-and-pull. For Lena, the feelings of community she experiences on Instagram can help her to push through her participatory reluctance (Cassidy, 2018), remaining on the app even when not particularly enjoying it: 'I don't always enjoy it, but I feel like I am very connected through Instagram.' For others, like we saw earlier with Rita, the rich informational experience she gains on Twitter overrides her desire to leave a platform she openly recognizes as toxic.

While all everyday engagements with social media can evoke particular intensities, the intense digital and affective labour required by feminist activist practices (Mendes, 2022) seems to carry particular pressures and reverberations. While interviewees generically referred to social media, and particularly TikTok, as addictive, they face special pressures as activists. Activist and feminist work tends to be framed as a 'labour of passion', which evokes various feelings of individual or collective political responsibility (Caldeira et al, 2025). The ambiences of the social media platforms where these practices occur can play an important role in exacerbating these pressures. Each platform's real and imagined audiences can reinforce feelings of responsibility towards vocalizing one's political views. As Lena exemplifies, these pressures can derive from being surrounded by activists as well as being isolated as the only activist in her social group: 'In non-activist circles, I end up being seen as "Lena, the activist". So, often they come up to me asking, "What about your opinion about that?", "You didn't share anything about this or that?". This creates more pressure to share than in a healthy activist environment.'

These active or implicit requests can create expectations for platforms like Instagram to be continuously used as an informative environment, either consuming content to keep themselves up to date or sharing content to inform others. For interviewees like Mourana, there could be an almost moral undertone to these pressures, as she felt that every moment she was not sharing, 'an evil continued to be perpetuated'. Again, the activist ethos accentuates the 'infra-structures' of feeling that characterize the social media connection, as Coleman (2018) described it, and conflicts with the particular moods activists go through.

The affordances and cultural expectations of 'response-ability' in platforms, the expectation that people should respond or react to every incoming interaction, can also contribute to feelings of pressure and saturation (Lupinacci, 2021, p 284), or anxiety and stress, as found among journalists interviewed by Bossio and Holton (2021). As Catarina put it, her attempts to decompress and temporarily distance herself from Instagram can be complicated by this: 'I'm always having to respond to messages. And then, whenever I spend a day without going there, I regret it. Because afterwards I have 300 messages to respond to.' In addition, the technological and commercial infrastructures of social media platforms ruled by logics of popularity and algorithmic promotion (van Dijck and Poell, 2013) also create expectations that people should strive for constant use and ever-growing platform metrics, which in these activist contexts can become problematically conflated with political legitimacy. Rita experienced these pressures, chastising herself: 'I'm like, "Rita, you created a page and now you don't post anything for a month".'

These dominant cultures of use, expectations and pressures highlight the time and resource-intensive nature of social media political uses, which do not necessarily correlate with effective political or social change (Edwards et al, 2020). As we expanded elsewhere (Caldeira et al, 2025), this can create feelings of tiredness, fatigue, exhaustion or even activist burnout (Gorski and Chen, 2015) that frame the affective experiences of these platforms. As platforms' stimuli and feelings of political responsibility coalesce, these can lead interviewees who see their social media activity as a 'civic duty', as Miguel put it, to neglect their own self-care in pursuit of these practices:

> I feel that I'm involved in so many things, that I need to reach everywhere, that I'm constantly failing, isn't it? Sometimes it's exhausting ... Sometimes I think: 'Ok, should I sleep one hour less and produce stories to reach 500, 700 people?' ... I feel like sometimes I neglect my self-care. (Mourana)

The dominant logic of contemporary feminisms as 'digital by default' thus relies not only on the expanding communicative possibilities that feminists could be taking advantage of on social media, but on the normative assumption that they *should* be actively engaging with these platforms (Fotopoulou, 2016, p 1000).

In order to deal with the pressures and shortcomings of digital activisms, our participants often reported disconnecting through what we have called 'small acts of disengagement' (Caldeira et al, 2025, adapting Picone et al's, 2019, small acts of engagement). These practices required less investment than more radical forms of disconnection (for example, boycotting platforms) and allowed participants to negotiate dis/connection flows in their everyday life. Some interviewees chose to disconnect by turning off their devices in certain situations (Rita), putting their phones out of reach (Ana Fidalgo, Lena, Telmo) or turning off notifications (Ana Sofia, Catarina).

However, to others, disconnecting did not always mean limiting access or time spent on digital devices. In our participants' accounts, moods were often entangled with their disconnection practices. For instance, when Carolina Fernandes feels saturated by political content online, she often turns to other sources of distraction or entertainment, like books, movies, or scrolling on TikTok, which she uses to 'unwind' and consume content that 'doesn't make her think much', similarly to Lupinacci's (2021) participants. Rita shared similar perceptions about using TikTok as a form of disconnection from feminist pressures, offering a refuge of 'mindless' digital activity: 'I would rather waste my time and be distracted on TikTok, making myself laugh, than constantly refresh my Twitter, or my Instagram or Facebook comments.'

Similarly, Carolina F. and Rita Oliveira deliberately avoided platforms perceived as more aggressive and hostile. Twitter was the most cited by participants as a 'toxic' environment.

> I avoid Twitter. Twitter melted my mental health, to the point that I decided: 'look, I can't do this anymore'. The gain I had was not worth losing my time or losing my sanity. So, I decided Twitter is not a good place for me. I prefer to do this [digital activism] on Instagram and YouTube. (Bruno)

Shifting from one platform to another deemed less problematic illustrates the fluidity of disconnection practices, as does disengaging from specific elements within platforms (Light, 2014). Many interviewees (for example, Catarina, Catarina Silva) saw comments as a confrontational space across platforms, causing some to refrain from commenting. Carolina F. often avoids commenting because it makes her uncomfortable and 'cringy'. To Marta, commenting is an unproductive experience:

> It's a waste of time. I think if it were a face-to-face conversation, it could be more prolific than a comment box. I think it is quite difficult to have a rich discussion in the comments on topics like those that I follow or interact with. Like, sometimes I think: 'if I start talking,

I'm going to have to hang out here on my phone for a day', and I don't have time for that [laughs].

In light of these tensions, as in other social movements (Chu and Yeo, 2020, on Hong Kong young activists), participants often elevated offline spaces as both more affectively valuable than social media environments and of greater symbolic weight for political and social transformation. Offline feminist spaces were described by some participants as more intimate and authentic, allowing for the creation of deeper 'human' connections. For instance, Marisa emphasized the importance of 'hugs' and 'looking people in the eye' as a fundamental part of her feminist activist engagements. As expanded in another publication (Caldeira et al, forthcoming), events like the 8M march could provide spaces for these intimate interpersonal engagements, with our observations documenting a pervasive sense of shared enthusiasm among the crowd and, at times, an almost party-like atmosphere – punctuated by laughter, drinks and live drums. However, as seen earlier, these offline spaces were not impervious to digital presence. If during the preparation of the march social media served informational and utilitarian uses, in the course of the march smartphones served to facilitate and at times document affectively charged encounters with others. As such, despite considering the online environment as more 'superficial', interviewees still widely recognized the importance of social media platforms as practical and logistical tools to make collective organization easier.

Conclusion

As platformized feminisms continue permeating everyday life, both off- and online, focusing on these affectively intense contexts (Mendes, 2022) allowed this chapter to highlight how social media ambiences bring together material, affective and situational elements to shape users' personal and political experiences. As our multi-sited research hopefully showed, the ways in which people live with and affectively experience feminisms on social media are highly situated, transient and frequently ambivalent – affected by the materiality and *vibes* of different platforms, contingent on platforms' communities, but also shaped by, and in turn shaping, the moods of users. The ambiences of particular social media platforms also respond to local cultures and communities of use, the affective impacts of their use depending on whether they are seen as facilitating community-building that goes beyond isolated conservative geographies or, conversely, invite hostile reactions from more traditionalist user bases. This chapter explored the complexities of navigating social media environments where the promise of political potential coexists with emotionally exhausting experiences or even overt anti-feminist toxicity – a tension that can be felt particularly in conservative and, increasingly, right-leaning contexts such as Portugal. These complexities create conflicting desires to remain connected, while simultaneously exerting feelings of reluctancy and unease that lead to efforts to at least temporarily or selectively push against particular platforms or particular elements within them. In these contexts, connection and disconnection are always in flux, particularly when the situational pull of platforms asserts itself given the need to mobilize for the next event, the next International Women's Day march, or in response to the next feminist debate.

Note

[1] More recent examples can include news of Meta's move against algorithmically recommending political content on its platforms (Leaver, 2024) or Mark Zuckerberg's latest 'toxic revamp' of his platforms (Morgan, 2025) that similarly stifled perceptions of their political potential.

References

Andiloro, A. (2023) 'Understanding genre as atmospheric assemblage: the case of videogames', *Television & New Media*, 24(5), pp 559–570.

Banet-Weiser, S. (2018) *Empowered: Popular Feminism and Popular Misogyny*. Duke University Press.

Barbala, A. M. (2024) 'The platformization of feminism: the tensions of domesticating Instagram for activist projects', *New Media & Society*, 26(10), pp 5802–5818.

Bengtsson, S. and Johansson, S. (2022) 'The meanings of social media use in everyday life: filling empty slots, everyday transformations, and mood management', *Social Media + Society*, 8(4). https://doi.org/10.1177/20563051221130292

Bossio, D. and Holton, A. E. (2021) 'Burning out and turning off: journalists' disconnection strategies on social media', *Journalism*, 22(10), pp 2475–2492.

Brown, M.-G., Carah, N., Robards, B., Dobson, A., Rangiah, L. and De Lazzari, C. (2024) 'No targets, just vibes: tuned advertising and the algorithmic flow of social media', *Social Media + Society*, 10(1). https://doi.org/10.1177/20563051241234691

Bucher, T. (2017) 'The algorithmic imaginary: exploring the ordinary affects of Facebook algorithms', *Information, Communication & Society*, 20(1), pp 30–44.

Caldeira, S. P. (2024) 'Instagrammable feminisms: aesthetics and attention-seeking strategies on Portuguese feminist Instagram', *Convergence*, 30(5), pp 1832–1848.

Caldeira, S. P. and Machado, A. F. (2023) 'The red lipstick movement: exploring #vermelhoembelem and feminist hashtag movements in the context of the rise of far-right populism in Portugal', *Feminist Media Studies*, 23(8), pp 4252–4268.

Caldeira, S. P., Jorge, A. and Kubrusly, A. (2025) '"How can you be a feminist if you're always online?" Online activisms, ambivalence, and dis/connection', *Television & New Media*, 26(4), pp 399–420.

Caldeira, S. P., Lamartine, C. and Kubrusly, A. (forthcoming) '"We go online to call them to the street": multidirectional flows in hybrid digital feminisms', *Signs: Journal of Women in Culture and Society*.

Campos, R., Simões, J. A. and Pereira, I. (2018) 'Digital media, youth practices and representations of recent activism in Portugal', *Communications*, 43(4), pp 489–507.

Cassidy, E. (2018) *Gay Men, Identity and Social Media: A Culture of Participatory Reluctance*. Routledge.

Cerqueira, C., Taborda, C. and Pereira, A. S. (2023) '#MeToo em Portugal: uma análise temática do movimento através de artigos de opinião', *Cuadernos.info*, (55), 1–21.

Chu, T. H. and Yeo, T. E. D. (2020) 'Rethinking mediated political engagement: social media ambivalence and disconnective practices of politically active youths in Hong Kong', *Chinese Journal of Communication*, 13(2), pp 148–164.

Clark-Parsons, R. (2018) 'Building a digital Girl Army: the cultivation of feminist safe spaces online', *New Media & Society*, 20(6), pp 2125–2144.

Coleman, R. (2018) 'Theorizing the present: digital media, pre-emergence and infra-structures of feeling', *Cultural Studies*, 32(4), pp 600–622.

Dean, J. (2005) 'Communicative capitalism: circulation and the foreclosure of politics', *Cultural Politics*, 1(1), pp 51–74.

Duffy, B. E. and Hund, E. (2019) 'Gendered visibility on social media: navigating Instagram's authenticity bind', *International Journal of Communication*, 13, p 20.

Dumitrica, D. and Hockin-Boyers, H. (2023) 'Slideshow activism on Instagram: constructing the political activist subject', *Information, Communication & Society*, 26(16), pp 3318–3336.

Edwards, L., Philip, F. and Gerrard, Y. (2020) 'Communicating feminist politics? The double-edged sword of using social media in a feminist organisation', *Feminist Media Studies*, 20(5), pp 605–622.

Fernández-Romero, D. and Sánchez-Duarte, J. M. (2019) 'Alianzas y resistencias feministas en Facebook para la convocatoria del 8M en España', *Convergencia Revista de Ciencias Sociales*, 26(81). https://doi.org/10.29101/crcs.v26i81.11943

Fotopoulou, A. (2016) 'Digital and networked by default? Women's organisations and the social imaginary of networked feminism', *New Media & Society*, 18(6), pp 989–1005.

Gago, V. (2020) *Feminist International: How to Change Everything*. Verso Books.

Garraio, J., Santos, S. J., Amaral, I. and Carvalho, A. S. (2020) The Unimaginable Rapist and the Backlash Against #MeToo in Portugal. Available at: https://www.europenowjournal.org/2020/03/09/the-unimaginable-rapist-and-the-backlash-against-metoo-in-portugal/

Gaskell, G. (2006) 'Individual and group interviewing', in Bauer, M.W. and Gaskell, G. (eds), *Qualitative Researching with Text, Image and Sound: A Practical Handbook*. SAGE, pp 38–56.

Gibbs, M., Meese, J., Arnold, M., Nansen, B. and Carter, M. (2015) '# Funeral and Instagram: death, social media, and platform vernacular', *Information, Communication & Society*, 18(3), pp 255–268.

Gorski, P. C. and Chen, C. (2015) '"Frayed all over": the causes and consequences of activist burnout among social justice education activists', *Educational Studies*, 51(5), pp 385–405.

Kanai, A. (2021) 'Intersectionality in digital feminist knowledge cultures: the practices and politics of a travelling theory', *Feminist Theory*, 22(4), pp 518–535.

Kanai, A. and McGrane, C. (2021) 'Feminist filter bubbles: ambivalence, vigilance and labour', *Information, Communication & Society*, 24(15), pp 2307–2322.

Karppi, T., Chia, A. and Jorge, A. (2021) 'In the mood for disconnection', *Convergence*, 27(6), pp 1599–1614.

Keller, J. (2019) '"Oh, she's a Tumblr feminist": exploring the platform vernacular of girls' social media feminisms', *Social Media + Society*, 5(3). https://doi.org/10.1177/2056305119867442

Kender, K. (2022) 'Tumblr is queer and Twitter is toxic: speculating about the vibe of social media spaces', in *NordiCHI '22: Nordic Human-Computer Interaction Conference*, ACM, pp 1–8. https://doi.org/10.1145/3546155.3547279

Khilnani, S. (2021) 'Sensations and solidarity: affect, ambience, and politics in digital literary narratives', *NECSUS European Journal of Media Studies*, 10(1), pp 173–193.

Kolehmainen, M., and Mäkinen, K. (2021) 'Affective labour of creating atmospheres'. *European Journal of Cultural Studies*, 24(2), pp 448–463.

Lamartine, C. and Cerqueira, C. (2023) 'Communicating through cyberfeminism: communication strategies for the construction of the international feminist strike in Portugal', *Social Sciences*, 12(9), p 473.

Leaver, T. (2024) 'Instagram and Threads are limiting political content. This is terrible for democracy', *The Conversation*. Available at: http://theconversation.com/instagram-and-threads-are-limiting-political-content-this-is-terrible-for-democracy-226756

Legard, R., Keegan, J. and Ward, K. (2003). 'In-depth interviews', in Ritchie, J. and Lewis, J. (eds), *Qualitative Research Practice: A Guide for Social Science Students and Researchers*. SAGE, pp 139–168.

Light, B. (2014) *Disconnecting with Social Networking Sites*. Palgrave Macmillan UK.

Lim, W. M., Jee, T. W., Loh, K. S. and Chai, E. G. C. F. (2020) 'Ambience and social interaction effects on customer patronage of traditional coffeehouses: insights from *kopitiams*', *Journal of Hospitality Marketing & Management*, 29(2), pp 182–201.

Lomborg, S. (2011) 'Social media as communicative genres', *MedieKultur: Journal of Media and Communication Research*, 27(51), pp 55–71.

Lupinacci, L. (2021) ' "Absentmindedly scrolling through nothing": liveness and compulsory continuous connectedness in social media', *Media, Culture & Society*, 43(2), pp 273–290.

Lupinacci, L. (2025) 'Mixed feelings: the platformisation of moods and vibes', *AoIR Selected Papers of Internet Research*. https://doi.org/10.5210/spir.v2024i0.13992

Markham, A. N. (2021) 'Echolocation as theory of digital sociality', *Convergence*, 27(6), pp 1558–1570.

Marôpo, L., Torres da Silva, M. and Magalhães, M. (2017) 'Feminismo online em Portugal: um mapeamento do ativismo no Facebook', in Pereira, S. and Pinto, M. (eds) *Literacia, Media e Cidadania - Livro de Atas do 4.º Congresso*. CECS, pp 280–293.

Massanari, A. (2017) '#Gamergate and The Fappening: how Reddit's algorithm, governance, and culture support toxic technocultures', *New Media & Society*, 19(3), pp 329–346.

Mattoni, A. (2017) 'A situated understanding of digital technologies in social movements. Media ecology and media practice approaches', *Social Movement Studies*, 16(4), pp 494–505.

Mattoni, A. (2020) 'Practicing media—mediating practice | a media-in-practices approach to investigate the nexus between digital media and activists' daily political engagement', *International Journal of Communication*, 14, p 18.

Mendes, K. (2022) 'Digital feminist labour: the immaterial, aspirational and affective labour of feminist activists and fempreneurs', *Women's History Review*, 31(4), pp 693–712.

Merrill, J. B. and Oremus, W. (2021) 'Five points for anger, one for a "like": how Facebook's formula fostered rage and misinformation', *The Washington Post*. Available from: www.washingtonpost.com/technology/2021/10/26/facebook-angry-emoji-algorithm/

Milan, S. (2015) 'Mobilizing in times of social media: from a politics of identity to a politics of visibility', in Dencik, L. and Leistert, O. (eds) *Critical Perspectives on Social Media and Protest: Between Control and Emancipation*. Rowman & Littlefield, pp 53–71. Available at: http://www.ssrn.com/abstract=2880402

Morgan, A. (2025) 'Why does Mark Zuckerberg want more "masculine energy" in the corporate world? The patriarchy is still in charge', *The Conversation*. Available at: http://theconversation.com/why-does-mark-zuckerberg-want-more-masculine-energy-in-the-corporate-world-the-patriarchy-is-still-in-charge-248600

Munro, E. (2013) 'Feminism: a fourth wave?', *Political Insight*, 4(2), pp 22–25.

Özkula, S. M. (2021) 'What is digital activism anyway? Social constructions of the "digital" in contemporary activism', *Journal of Digital Social Research*, 3(3), pp 60–84.

Paasonen, S. (2021) *Dependent, Distracted, Bored: Affective Formations in Networked Media*. MIT Press.

Papacharissi, Z. (2009) 'The virtual geographies of social networks: a comparative analysis of Facebook, LinkedIn and ASmallWorld', *New Media & Society*, 11(1–2), pp 199–220.

Picone, I., Kleut, J., Pavlíčková, T., Romic, B., Møller Hartley, J. and De Ridder, S. (2019) 'Small acts of engagement: reconnecting productive audience practices with everyday agency', *New Media & Society*, 21(9), pp 2010–2028.

Ponder, M. L., Addie, Y. O., Meux, A. I., Tindall, N. T. J. and Gulledge, B. (2023) 'Does online activism impact offline impact? A cultural examination of slacktivism, "popcorn activism," power, and fragility', in Wallace, A. A. and Luttrell, R. (eds) *Strategic Social Media as Activism*. Routledge, pp 257–277.

Pruchniewska, U. (2019) *Everyday feminism in the digital era: gender, the fourth wave, and social media affordances*. Doctoral Thesis in Philosophy, Temple University.

Roquet, P. (2012) *Atmosphere as culture: ambient media and postindustrial Japan*. Doctoral Thesis in Japanese Language and Film Studies, University of California, Berkeley.

Sampson, T., Maddison, S. and Ellis, D. (2018) *Affect and Social Media: Emotion, Mediation, Anxiety and Contagion*. Rowman & Littlefield.

Santos, A. C. and Pieri, M. (2019) 'My body, my rules? Self-determination and feminist collective action in Southern Europe', in Flesher Fominaya, C. (ed) *Routledge Handbook of Contemporary European Social Movements*. Routledge, pp 196–209.

Sørensen, A.S. (2009) 'Social media and personal blogging: textures, routes and patterns', *MedieKultur: Journal of media and communication research*, 25(47). https://doi.org/10.7146/mediekultur.v25i47.1698

Şot, İ. (2022) 'Fostering intimacy on TikTok: a platform that "listens" and "creates a safe space"', *Media, Culture & Society*, 44(8), pp 1490–1507.

Tucker, I. M. and Goodings, L. (2017) 'Digital atmospheres: affective practices of care in Elefriends', *Sociology of Health & Illness*, 39(4), pp 629–642.

Usher, B. (2020) 'Rethinking microcelebrity: key points in practice, performance and purpose', *Celebrity Studies*, 11(2), pp 171–188.

Van Der Wal, A., Valkenburg, P. M. and Van Driel, I. I. (2024) 'In their own words: how adolescents use social media and how it affects them', *Social Media + Society*, 10(2). https://doi.org/10.1177/20563051241248591

Van Dijck, J. and Poell, T. (2013) 'Understanding social media logic', *Media and Communication*, 1(1), pp 2–14.

Van Zoonen, L. (2011) 'The rise and fall of online feminism', in Christensen, M., Jansson, A., and Christensen, C. (eds) *Online Territories: Globalization, mediated practice, and social space*. Peter Lang, pp 132–146.

THREE

Affective Temporalities in Pilgrimage: Anticipation, Presence and (Pro)longing

Ana Jorge, Filipa Neto, Ana Kubrusly and Edna Santos

Introduction

While interest in religion decreases in contemporary societies, interest in spirituality, meditation and holistic wellness grows (Heelas and Wollhead, 2005). Pilgrimage can be broadly defined as travelling from one's home to a sacred place, in which the symbolic act of walking gains immense significance, over the destination itself (Slavin, 2003; Vilaça, 2008). In its post-secular form, pilgrimage combines religion, or spirituality more broadly, with tourism and self-development (Nilsson and Tesfahuney, 2018; Beckstead, 2021). It intersects, and partly overlaps, with self-help culture and popular psychology to which the metaphors of *path* and *journey* are so dear, connected to ideas of self-transformation and spiritual self-discovery which require a voluntary, 'partial or full withdrawal from everyday challenges' towards a good life (Nehring et al, 2016, p 90).

Despite – or precisely because of – this transformation, pilgrimage remains relevant across cultures and religions, and it is one of the forms that best reflect present-day religious rituality (Vilaça, 2008). It is an 'embodied, emotional-affective, sensory, and aesthetic experience' (Maddrell and della Dora, 2013, p 1123). Crucially, pilgrimage also has relational and social dimensions, to which anthropologists Turner and Turner (1978) referred as *communitas*, the communion with people who share similar values and goals during a religious experience, suspending everyday social differences. Relevant here is Berlant's (2008) understanding of *communitas* as the affective intensity of a sense of belonging with others, regardless of the realm in which it occurs. Particularly, since it involves human gatherings, pilgrimage evokes *effervescence*. In the early 20th century, Durkheim (2001 [1912]; Serazio, 2013) drew on the parallel between religious gatherings and other convergences in social life to explore the sharing and elevating of social sentiments.[1]

Digital media play a significant role in pilgrimages across religions and various parts of the world (Caidi, 2023; Wright-Ríos and Martínez-Don, 2024; Hussain and Wang, 2024). In earlier work, one of the authors of this chapter (Jorge, 2023) explored how, in the aftermath of COVID-19, pilgrims used digital media during their routes to Fátima and Santiago de Compostela, and how they managed dis/connection. Pilgrimage appears as an escape from the 'always-on' culture (Mascheroni and Vincent, 2016) and the accelerated pace of society, to find clarity and evolve as an individual (Nilsson, 2018). Therefore, there is an expectation that pilgrims interrupt their habitual use of digital media during the pilgrimage, similar to what happens in active or nature tourism (Tosoni and Turrini, 2018; Schwarzenegger and Lohmeier, 2021).

While pilgrimage partly operates on an alternative economy, it is not only heavily promoted by tourism offices but also deeply entangled with digital media (Jansson, 2018; 2024), including by extensive social media representation generated by pilgrims and use of booking platforms and maps (Jorge, 2023). Furthermore, pilgrims seek to remain in contact with their significant ones, by selectively using time, services and contacts (Dickinson et al, 2016; Rosenberg, 2019).

Despite being seen as transitional and the 'quintessential liminal activity' (Beckstead, 2021, p 85), an activity that breaks a boundary and suspends part of the functioning of everyday life, pilgrimage confirms the latter in novel ways (Maddrell and della Dora, 2013) and the transformations they spark are enduring (Beckstead, 2021). Crucially, the entanglement of pilgrimage and digital media, evidence of a post-digital age, is one way in which this liminality is challenged (Jorge, 2023), and thus appears as a rich locus for interrogation about digital media dis/engagement.

In seeking to understand how digital media atmospheres and pilgrims interact, temporalities become paramount. As one of the oldest forms of human mobility, pilgrimages are connected to historical tradition and use ancient, slow forms of transport (walking, horseback riding, biking) in a high-speed society (Howard, 2012; Wajcman, 2015). Consequently, pilgrimage resonates with notions of presentism and mindfulness that inform popular discourses of disconnection (Syvertsen and Enli, 2020), just as it does with transcendence and spirituality.

This chapter examines how temporally driven collective feelings are enacted through digital media in relation to pilgrimage, involving not only mobility, spatiality and embodiment but also relationality and morality. Lefebvre affirms that 'everywhere where there is interaction between a place, a time, and an expenditure of energy, there is rhythm' (2004, p 15). We particularly draw on the notion of *affective temporalities* offered by Nikunen (2023) to understand how emotions are mobilized through time. We address pilgrimage as a digital affect culture composed of flows of emotion that contain both resonance and contagion (Hitchen, 2021), and disalignment and polyphony (Döveling et al, 2018; Ural, 2023). Respectively, through the themes of *anticipation, presence and (pro)longing*, we explore how different social media platforms and other digital services interact with pilgrims to build *communitas*, augment effervescence and process the emotion of longing by enabling various forms of feeling time.

Our analysis also brings out how multiple temporal flows converge in pilgrimage, and contain ambivalence (Sharma, 2014; Firth et al, 2020; Germann Molz and Buda, 2022).

Study

We explore Christian pilgrimage cases close to the Portuguese reality, yet internationally significant: Santiago de Compostela, in neighbouring Galicia, Spain; traditional sanctuary Fátima, in the centre of mainland Portugal; and World Youth Day (WYD), which took place in Lisbon in 2023.

The Camino de Santiago is an international network of pilgrim routes dating back to medieval times (Nilsson and Tesfahuney, 2018) that connects various parts of Spain, France and Portugal to Santiago de Compostela; it is a site of pilgrimage all year round and, since becoming a UNESCO World Heritage Site in 1985, is promoted by tourism institutions, which illustrates its growing secularization (Vilaça, 2010). It is 'now a sacred space mostly for individual spirituality rather than organized religion' (Nilsson, 2018, p 22). Camino de Santiago is known for the 'inter-subjective experiences, spaces of being and sharing life-experiences' (Nilsson and Tesfahuney, 2018, p 168). In 2023, Santiago certified 446,000 pilgrims (in other words, those who had walked at least 100 km or travelled at least 200 km by horse or bicycle – Pilgrim's Welcome Office, 2025), more than half of whom were foreigners to Spain. Portugal has, since 2015, invested in various paths towards Santiago; these paths were, by 2023, among the most popular routes taken by the growing number of international pilgrims (Santos, 2024).

Fátima is a Marian sanctuary with biannual celebrations, held on the evenings before 13 May and 13 October, commemorating Mary's apparitions to three young shepherds in 1917. Pilgrims come mainly from the Portuguese mainland, but Catholics from many parts of the world go to the sanctuary. In 2023, Fátima received 6.8 million pilgrims (1.1 million of whom were pilgrims to WYD; Sanctuary of Fátima, 2024). The pilgrimage to Fátima is imbued with some degree of effort or resilience (Sharma and Timothy, 2023), from sacrifice (Egan, 2010) to 'penance, reverence, ... petition' (Maddrell and della Dora, 2013, p 1108) or thankfulness. This explains its association with a form of popular religiosity (Vilaça, 2010). Pilgrims usually walk to Fátima in organized groups, walking on the sides of the roads and departing from their homes or parishes. Praying is done at various points along the journey and people sleep in lodges or schools en route. Over the last few years, there has been a heavy investment in Caminhos de Fátima as a tourist infrastructure, as part of 'tourism of faith' (Caminhos da Fé, 2021); part of this route is the same as towards Santiago, so it is common to find the symbols of both destinations together. Rota Carmelita, one of those paths, in central Portugal, is a route that directly evokes the religious order of the longest-surviving shepherdess, Irmã Lúcia.

World Youth Day is a worldwide Catholic gathering between young people and the Pope celebrated approximately every three years in a different host country since 1984. In 2018, it was confirmed that Lisbon would be the host city for the following WYD, which eventually took place between 1 and 6 August 2023, with an estimated 1.2 million pilgrims from 150 countries (Lusa, 2023). Given the volume of pilgrims and the fact that the target group has limited financial resources for travel, accommodation is organized through schools, public spaces, or domestic volunteer hosts (Gonzalez et al, 2019). The event included religious meetings and gatherings, as well as the 'youth festival' including concerts, conferences, cultural (theatre and dance, museums, cinema, religious exhibits) and sports activities (Fundação JMJ Lisboa, 2023). This programming, similar to other WYDs, explains the approximation to festivals and the label of 'Catholic Coachella' (Bogacz-Wojtanowska et al, 2019).

Our data collection, in 2023, started with an exploration to understand which social media platforms and digital spaces (maps, apps, repositories) were mostly used in each pilgrimage case; then, we identified popular Instagram hashtags and key institutional accounts, Facebook groups and pages, and one TikTok hashtag. We scraped data directly related to the Santiago and Fátima pilgrimages and the WYD event (before and after). We sampled the extensive corpus to one third, in systematic sampling across the period of collection – the final corpus is described in Table 3.1.

Furthermore, we analysed a total of 40 interviews with participants in the pilgrimages. In 2023, we conducted interviews with 16 WYD pilgrims, volunteers, organizers or hosts (several of whom held or shifted between these roles), adding to 13 interviews with pilgrims to Fátima and Santiago conducted in 2021 for our previous study (Jorge, 2023). Interviews focused on the motivations for pilgrimage (or involvement as organizer or host) and the uses and perceptions of digital media. WYD interviewees were recruited with an e-flyer circulated on the research team's social media, and through direct invitations to two TikTok account owners that were identified during social media scraping (though another 11 from TikTok and Instagram were invited but did not respond). Participants in 2023 were compensated with a voucher, and interviews had an average duration of 54 minutes. All but one of the interviews took place over videocall, at their preference; however, only audio was saved.

Additionally, in 2023 we did observations on the ground: one of the authors did an ethnographic study on the Portuguese Camino de Santiago in the spring of that year, including three interviews with hostel hosts, and eight informal interviews with pilgrims; and two of the authors visited three hostels on the route between Lisbon and Fátima in the autumn of the same year. Observations included exemplary photos, sketches of the venues, and notes on spaces, interactions and experiences, focusing on uses of digital media and places. All interviews were transcribed verbatim.

All data were saved on external drives only accessible to the authors. The corpus was formed from publicly available online

Table 3.1: Digital media corpus

Digital media	Santiago	Fátima	WYD
Instagram			
Hashtags	3 (#caminhode santiagode compostela – 839 posts; #caminode santiago – 945 posts; #santiagode compostela – 898 posts)	2 (#fátima 2023 – 212 posts; #caminhodefatima – 609 posts)	3 (#estamosa caminho – 185 posts; #jmj – 439 posts; #lisboa 2023 – 110 posts)
Highlights			3 (lisboa 2023_en – 199 stories; lisboa 2023_pt – 445 stories; cop_jmj_lumiar – 43 stories)
Locations		3 (Caminhos de Fátima – 13 posts; Santuário de Fátima – 50 posts; Santuário de Fátima, Portugal – 50 posts)	
Facebook			
Groups	2 (66 posts)	2 (310 posts)	2 (341 posts)
TikTok			#jmj2023 (374 videos)
Giphy	168 GIFs		24 GIFs
Maps	6 (Google Maps, Wikiloc, Komoot, Vagamundos, Forum, Google Earth)	3 (Google Maps, official website, Wikiloc)	3 (Google Maps, official website, Waze)
Apps	1 (Buen Camino Santiago)	2 (Caminhos de Fátima, Santuário de Fátima myEyes)	6 (Lisboa 2023, WYD Global Race, Passo-a-Rezar, JMJ Lisboa 2023 Córdoba, WYD 2023 LISBIN Magnificat, Dias nas Dioceses)

data, and given the volume of material, consent was not sought. We followed the Association of Internet Researchers' (AoIR) *Ethical Guidelines* (franzke et al, 2020) and considered the material anonymously; examples are anonymized. Consent was obtained from the interviewees, and the choice of identification or pseudonym was offered to WYD pilgrims; interviews to hosts of *albergues* were anonymized; and anonymized informal interviews with Santiago pilgrims were also conducted after the researcher disclosed their role, of which only notes were taken at a later time. The codebook for the materials was developed inductively and deductively, after familiarization with the data and socialization among the team. For online data, we created categories that were attentive to affordances and platforms, and the analysis was multimodal (Bouvier and Rasmussen, 2022), and conducted on Excel; interviews and observation of pilgrimage routes were thematically analysed using MaxQDA software. In terms of positionality, the researcher who collected the data for Camino de Santiago is a Catholic person walking the pilgrim trail for the first time; one author is Catholic and participated in WYD in Lisbon; one is a non-practising Catholic; and one is agnostic, without formal ties to the Catholic church. These different positions translated into different levels of knowledge about, and distancing from, the object under study.

Anticipation: excitement and apprehension

The pilgrimage starts before the physical journey takes place, or, as a comment on a Facebook group dedicated to Santiago reads, 'The fact that you want to walk the Camino means that you are already doing it.' Preparing for pilgrimage involves anticipatory work that is as much affective as logistical. On the one hand, anticipation can be seen as a 'premeditation that mobilises affect' (Nikunen, 2023, p 178), or as 'the temporal and affective structuring of intimacies, socialities and relationships that platforms enable and engender' (Koivunen et al, 2024, p 2). On the other, as other travellers for leisure or work (Tosoni and Turrini, 2018), pilgrims explore the logistical affordances of digital media (Jansson, 2024). The pilgrim-to-be acknowledges a motivation to participate in pilgrimage, which may involve apprehension about moving and facing something challenging or, indeed, unknown (Sebald, 2020; Ekmekcioglu et al, 2023). As this uncertainty is dismantled through practical organization, an important sense of unity, belonging and community is also created. Social media appears as a central space through which anticipation is cultivated (Nikunen, 2023).

For instance, for WYD, TikTok video 'This summer WE WILL SEE EACH OTHER AT THE ✝WYD' (original in Spanish) deployed lively music, young people at concerts, on buses and walking, projecting a vibrant and fun atmosphere, using archive images. Other, more 'serious', videos emphasized the values promoted by the church for the event: 'World Youth Day also seeks to promote peace, unity and fraternity among peoples and nations around the world'. This content sought to mobilize pilgrims, but crucially volunteers and hosts. Preparation for a global, mass event like WYD was also depicted on dedicated Instagram accounts by groups of young people from around the world, for months before the event, recounting their motivations to attend and social and religious events, including activities to raise funds to travel.

Social media features also work to build up the atmosphere for pilgrimage. Crucially, countdowns enacted through icons or posts fed anticipation: in the case of WYD, the number of days until the main event; or in the case of Santiago, the number of cities yet to walk through. Additionally, GIFs and stickers played a significant role in creating anticipation for the event, as they were promoted by the organizers and, in the case of Santiago, commercial entities within the industry surrounding the Camino. WYD organizers created GIFs featuring the event dates or location, along with a motto in the form of a hashtag: #weareonourway (original in Portuguese and Spanish). In the Santiago pilgrimage, similar formats are available; one GIF reads: 'I'm going from ★insert name of city★ to Santiago' (original in Spanish).

On Facebook, we found solidarity atmospheres where people sought and gave advice on preparing for the three pilgrimage cases. In Facebook groups dedicated to Fátima, the main topics were practical advice as well as pilgrims' previous experiences. This was especially relevant during the containment of the COVID-19 pandemic, when there was great uncertainty about whether it was possible to cross district or country borders (Jorge, 2023). Both in Fátima and Santiago groups, there is a strong emphasis on the symbolic parts of the pilgrimage, respectively, the pictures and statues of Mary, and the scallops and the yellow and blue arrows. Experienced pilgrims in the Camino offer advice to new pilgrims on their doubts and uncertainties about physical preparation, places to stay or the number of kilometres to walk per day. Practical advice goes hand in hand with mantra-like messages (for example, 'you never walk alone'), creating a feeling of *communitas* before starting to walk the Camino. A more personalized message reads: 'Buen Camino to all. The tricks: Drink water before you are thirsty. Rest before you are tired. Smile always. Say hello to everyone, those who walk and those who stay. Don't rush it.'

Waiting means affectively investing in something to happen, and embodying or ritualizing that wait (Sebald, 2020). The physical and practical aspects of pilgrimage are imagined, discussed and managed. To pilgrims going to Santiago, these aspects of their travel were crucial, which speaks of the increasingly post-secular nature of this pilgrimage. We found Instagram posts by pilgrims to Santiago depicting what they would carry in their backpacks, usually following the pattern of spreading all items on the floor for a still image. A similar trope was adapted on TikTok around WYD, as in the video 'What can we expect from WYD 2023?': girls showed the outfits they were planning on packing for the event; this was entangled with practical advice on how to pack for the expected heat in

Lisbon in August (loose clothing, hats, and so on) but also to adapt to religious spaces or ceremonies (for example, something to cover the shoulders if needed). This example illustrates how anticipating is not just logistical but also moral; it should not be seen as a coincidence that we found this type of video only by girls, pointing to a gendered phenomenon.

A more religious and spiritual crescendo was especially discernible for WYD and Fátima, because they are mass celebrations. Getting closer to the pilgrimage means preparing, in the form of gatherings, masses, prayers and reflections – which can be followed by parties – as we learned through Instagram posts from various groups. This spiritual preparation was also enabled by apps, such as Passo-a-Rezar (a pun between 'step' and 'getting started' praying), which has released a 10-minute reflection each day since November 2021 and promoted a monthly reflection inviting people to 'join, with their prayers, the preparation for the encounter'.[2] Portugal promoted Days in the Dioceses as WYD 'pre-events', as groups of young people from all over the world travelled around the country doing activities with the young people from those parishes. As one pilgrim described these gatherings on an Instagram post, 'the perfect preparation for what was to come'.

The fact that the build-up towards the pilgrimage is done amid apprehension was especially visible at WYD. As one of our interviewed pilgrims recalled: 'We were a month away from going to the WYD, and nobody knew anything. Everyone was very anxious because we did not know, we did not know where we were going to stay, we knew absolutely nothing.' (49-year-old, Portuguese woman). There was a strong contestation in the news media regarding the state's money invested in the religious event, and doubts about the ability to organize and accommodate a vast number of people. Two of our interviewees echoed this uncertainty when they told us, after the event, that they had feared the logistics of the event might fall through, and expressed personal anxieties about the crowds being overwhelming.

Presence: liminality and realtimeness

During the pilgrimages, the notion of presence appears in sometimes tensional ways: on the one hand, as a shared energy between people, places or with transcendence; on the other, as adhering to normative regimes of digital media culture for *presencing* (Couldry, 2012); that is, keep presenting oneself and engaging with the world through digital media.

Circulating norms indicate that pilgrims should not use those media at particular times or places, particularly given their religious, spiritual or self-invested nature. As was made clear in our interviews, many Santiago and Fátima pilgrims look at the opportunity of pilgrimage as a 'pause' in everyday life, including the expectations related to constant digital connection. Two young pilgrims to WYD told us they usually attend church and mass because those are a few of the places and moments where there is silence. Observing silence or chanting together in a crowd increases the sense of shared energy as effervescence, as both pilgrims to Fátima and WYD indicated. Moreover, disconnecting from digital media during the pilgrimage is seen as either a consequence (focusing attention on the ground) or as a prerequisite to engage with the experience (either with nature, oneself and/or others). Susan, an English female pilgrim, told us that she avoided using her smartphone while on the path so it would not impede her connection with other people; however, she would need to use it to look for services such as accommodations and places to eat.

In fact, as they perform the pilgrimage, individuals rely heavily on digital media as wayfarers, using maps, schedules and apps to navigate and engage with the experience (Pink and Hjorth, 2014). Pilgrims gather navigational information from apps (for example, Buen Camino de Santiago) or social media (for example, one Facebook group dedicated to the Camino de Santiago provides real-time information on weather, including fire alerts in summer). While the smartphone was considered the Swiss Army knife, incorporating several functions in a light portable device, by all pilgrims to Santiago or Fátima, some of them also used smartwatches, cameras and GoPros to assist their pilgrimage or its documentation. Quantifying the experience is relevant for some pilgrims (time, distances), which feeds some of the social media content.

Pilgrims also use social media to build on the feeling of *communitas* (Turner and Turner, 1978; Berlant, 2008). The interaction between pilgrims is a crucial aspect of their practices, as it contributes to the construction of a sense of recognition and sharing of feelings and experiences that are similar and common among those who undertake the pilgrimage (Collins-Kreiner, 2010). On Instagram, we found numerous posts featuring photos of groups of pilgrims to Santiago, celebrating meetings and reunions, sharing meals or taking moments of rest. 'Unity is strength', reads the caption of a post that shows the feet – wearing hiking boots – of a group of pilgrims on the Camino de Santiago. As other tourists (Edensor, 2016), pilgrims wish to navigate the routes in safe ways, but also to be surprised and for the experience to be spontaneous and complete with unexpected interactions.

Hosts of some of the *albergues* on the Camino promote receptions and meals to bring guests together and invite them to connect – discouraging the use of digital media during those moments. Architecture, decoration, as well as programming and written or unwritten rules create an atmosphere that favours conviviality over digital media use. The atmospheres also evoke the past, emphasizing local food, traditional crafts and frugality. These themes are echoed on many of the GIFs related to Santiago found on Giphy. Through this association with the past, pilgrims become part of a historical community.

Through mediated forms of popular religious celebrations, a sense of affective public can be created by both those who adhere to it and those who question and subvert its meaning (Papacharissi, 2014). On the one hand, technologies enable livestreams or simultaneous translations in support of synchronous and communal celebrations, across distance or language. For example, pilgrims from Guanarito, Venezuela, engaged with WYD at a distance: they organized their own version of the event, attended the main events with the Pope in Lisbon online in real-time, and shared the experience through Instagram. On the other hand, numerous memes teased political participants or fervent pilgrims at WYD. President Marcelo Rebelo de Sousa is the protagonist of several of these memes, starting with the overjoyed reaction to the confirmation, in 2019, of holding WYD in Lisbon – 'we hoped, we wished, we made it' – that has become part of the vernacular in the country; or for overenthusiastically saluting the Pope on arrival.

Besides photos with other pilgrims, many use social media as a diary of their journey. Some of our interviewees who had pilgrimed to Fátima and Santiago in 2021 told us that they would take some pictures on the journey but had to focus on the hard walking; they would, if at all, post at the end of the day, when resting at the lodge. We found abundant content under Santiago-related hashtags portraying moments from the day's walk, selfies and reflections on the journey, which range from personal to spiritual and/or religious. The intensity of the moment of arrival in Santiago de Compostela as the end of the pilgrimage also motivated some pilgrims to livestream their experience on Instagram (which we found as a video on their Instagram feed). On Facebook groups dedicated to the pilgrimage to Fátima, users will often share content as they are taking the pilgrimage or when reaching the sanctuary. On Instagram, one caption reads: 'Pilgriming to Fátima. Getting closer with each step!'; and another uses the hashtags '#firstdayofFátima2023 #onmywaytoMOTHER' (original in Portuguese).

This sense of urgency to post was more evident in content related to WYD, where much of the content seemed to have been posted instantaneously, and there were many live broadcasts. Many publications featured images of the crowds during the event, evoking a sense of belonging and unity, highlighting the importance of the gathering of people from different parts of the world united by a common feeling. The crowd, in all three pilgrimages analysed, exemplifies how belonging and resonance is felt, on the one hand, as an embodied experience of sameness and, on the other, as a phenomenon that can be bigger than the sum of its parts (Miller, 2015). From interviews with participants – even if not all were pilgrims – in WYD and Camino de Santiago, we got insights about how they felt the effervescence, the excitement and rush of living exceptionally intense and emotional moments in a massive crowd:

> I don't know, the feeling is awesome, I mean, wonderful. I don't know, I can't ... It's one of those things that you can't put into words, I could be here talking about happiness, joy, compassion, I don't know, but it's one of those things that I have a hard time expressing, even explaining to other people who haven't been there what the feeling is. (Portuguese pilgrim to WYD)

> In spiritual terms, it was completely different because the city itself seemed different, there was more spirit, there were more young people. (Portuguese host to WYD)

This feeling often *compels* some to share with others, especially non-pilgrims, what they are experiencing, whether on Instagram stories, WhatsApp pictures, or text, for instance. There is a ritual dimension (Couldry, 2003) especially to mega-events such as WYD but also Fátima celebrations, whereby participants feel not just part of a crowd, but of an event that the whole society is attuned to, even if they are not physically present. The urge to share acts as an 'infra-structure of feeling' (Coleman, 2018), an impulse provoked by the intensity of the experience as well as the culture of presentness and the built-in pressure for synchronicity orchestrated by both social media culture

(Jordheim and Ytreberg, 2021; Lupinacci, 2024) and devices (Weltevrede et al, 2014).

(Pro)longing: overflow and breaks

Pilgrims often use online spaces after their journeys to make sense of, share, remember and celebrate their experiences (Hussain and Wang, 2024), as well as take part in wider narratives and public discussions surrounding pilgrimage (Buitelaar, 2023). In this section, we examine how social media content by pilgrims in our studies reflects an overflow of emotion, building on specific emotions such as gratitude, nostalgia and *communitas*, and is enmeshed in wider discourses of division or subversion.

Pilgrim journeys are often portrayed online as transformative experiences of personal and spiritual growth, reinforcing the idea of pilgrimage as a rite of passage or a quest for meaning (Polus and Carr, 2021; Buitelaar, 2023). In a Facebook group dedicated to Fátima pilgrims, a woman wrote about the impacts of pilgrimage: 'You are renewed. You are also transformed. And you become another person through the feet that walked the path.' On an Instagram post, a woman shares that during her pilgrimage to Santiago: 'each second calms your soul, brings happiness to the heart, and makes us different!' A few Santiago pilgrims shared pictures of tattoos featuring common symbols – such as arrows and scallops – or sayings like *ultreia* or *buen camino*. WYD pilgrims also shared in interviews how their experience at the event was *indescribable*, unforgettable and, ultimately, transformative.

After their journey, many pilgrims shared expressions of gratitude for their experience, recognizing and praising the other parties involved. A pilgrim in a Fátima Facebook group posts her testimony after her pilgrimage with the status 'feeling grateful' and writes: 'Thank you for all those who crossed my path and to this group for all the help, thank you 🙏.' This quote illustrates how the pilgrim community is constructed both online and offline, with digital spaces allowing for asynchronous encounters between those who have been, are, or plan to go on this type of journey.

In the specific case of WYD, digital content posted after the event primarily focused on expressions of gratitude towards organizers, volunteers, sponsors and the city of Lisbon itself. Here, the community goes far beyond the pilgrims. On Google Maps, pilgrims left five-star reviews on the main location of the event, writing about the deep and wonderful spiritual experience they had. In our interviews, volunteers expressed feelings of *pride* towards the positive feedback about the event. A 22-year-old Portuguese female volunteer told us that everything that happened seemed 'inconceivable' before.

In this context, memory-making appears to be a central aspect of pilgrims' journeys, as was confirmed by our interviewees, who document their paths with photos and videos (Buitelaar, 2023). On digital platforms, we can recognize patterns across the content. For instance, in Santiago, pilgrims posed with open arms in front of the cathedral, symbolizing the end of their path and the peak of their pilgrimage experience. Others celebrated the anniversaries of their experience in Santiago and Fátima with Instagram posts, with a picture of themselves or the landscapes. This process of curating digital memories is a way for pilgrims to keep their experiences alive and give them meaning (Caidi et al, 2018), allowing them to relive the affective aspects of their journey. Reliving through photos and sharing enhances the remembrance of the pilgrimage; as one pilgrim wrote in an Instagram caption:

> I walked the Camino de Santiago in the summer of 2013. I always knew that I would go back one day, but now that 10 years have passed, and when I open my photos from that time, the sensations I felt at that moment seem to come alive. The endless fields, the wind, the sunshine, the flowers, the trees, the food and wine, and the people.

The nostalgic tone used to discuss post-pilgrimage experiences was evident in both our online and offline data collection. Nostalgia – as it is explored within social media – constitutes a type of performance, or narrative event, 'that connects time, space, and affective feeling' (Conner, 2023, p 2). Besides posts from earlier experiences of pilgrimage, what became salient in our material about WYD were the expressions of nostalgia immediately after the event. A photo shared on Instagram after WYD features a group of seven young people sitting in a circle on the dirt – at a camping site – smiling at the camera above them. In the caption and comments, heart emojis (♡) highlight how they miss struggling and going through the 'rough patches' of the event together. This illustrates how reminiscing might contain the entanglement of less pleasant aspects, as pointed out by our interviewees – being sleep-deprived, dealing with heat, crowded spaces – with the positive and exceptional parts of the pilgrimage experience.

Moreover, the stories and experiences shared by pilgrims online also play a role in the construction of a broader sense of community and identity (Caidi et al, 2018; Buitelaar, 2023). Shared 'digital memory spaces' can reinforce connections to shared past experiences and among the people engaging with them (Ekelund, 2022). In our data, we observed how this phenomenon occurred across various platforms, including Instagram posts, TikTok videos and Facebook groups. In interviews with WYD participants, the sentiment of unity among Catholic youth was highlighted as one of the main takeaways of the event. For instance, a 26-year-old female interviewee, who had previously received negative comments when sharing religious content on TikTok and Instagram, experienced a renewed sense of belonging and hope after participating in WYD.

However, this cohesive community among pilgrims is not without its own divisions, or tensions and disputes with ordinary society and other agents involved. Online, this manifests in conflicting narratives about pilgrims' behaviour, sanctioning what is considered against the prevailing norms. The complaints, by non-WYD participants, regarding security issues (visa policy not being strict enough), overburdened public transport and insufficient garbage collection made WYD participants feel that they had a duty to share with others their testimony of the event with a positive outlook. Instagram posts and Facebook comments, and broader media discussion questioned the legacy of WYD 2023, specifically the utility of the infrastructure built for it, as has happened in other mega-events (Bogacz-Wojtanowski et al, 2019). These divergent portrayals led some of our interviewees to feel uncomfortable with or even resentful about the negative aspects of the mega-event: a 24-year-old female pilgrim disliked the online reports about pilgrims getting drunk and partying in the city during the event; and a 17-year-old male pilgrim was upset at the trash pilgrims left behind in one of the main venues, Parque Tejo, after the event.

Across our data, it became apparent how nostalgia, gratitude and emotional engagement were linked to a sense of urgency for the next opportunity to participate in a pilgrimage. In the case of WYD, as soon as the next host city – Seoul – was announced, pilgrims across platforms said their thankful goodbyes to Lisbon, anticipating a future encounter in 2027, or even before, at the 2025 Jubilee in Rome. A Facebook group named 'JMJ [WYD] LISBOA 2023' changed its name to 'JMJ [WYD] SEUL 2027' right after Lisbon. Many of the WYD pilgrims we interviewed expressed that they were inclined to make a pilgrimage again, yet they were uncertain about logistics and finances in one year's time, let alone four. In the case of Fátima, a pilgrim shared pictures of herself and her companion on a Facebook group with a lengthy testimony about her positive experience, ending with 'and now that I am home, I want to go back ♥ [red heart]'. We found many instances of pilgrims wishing to return to Santiago, setting their schedule (for example, every year). One user posted: 'New goal to give new meaning to life! I really miss the Camino …

next year, God willing, we'll be there again 🙏 [praying] 😊 [beaming face with smiling eyes]'. These posts emphasize the transformational intensity of pilgrimage, and the willingness to solidify that transformation by repeating it.

Conclusion

This chapter demonstrated how pilgrims engage with social media to work through a desire to move on pilgrimage, building up expectations as much as reducing uncertainty. Digital media act as wayfarers for pilgrims (Pink and Hjorth, 2014), yet norms require that they disconnect and make themselves present to engage with themselves, others and/or the landscape, but also to evoke the historical heritage that the pilgrimage represents. On digital media, pilgrims also extend, process and relive the emotional intensity of the pilgrimage, including their feeling of belonging to a historical and/or cosmopolitan group. There is a cyclicity between remembrance and anticipation, that can be interpreted, following Lefebvre (2004), as cosmic; cyclical time does not transcend linear time, but rather counters flatness and creates a flow of emotions where social life is lived with a rhythm rather than a repetition.

Throughout, different layers are juxtaposed in an ambivalent, affective atmosphere of pilgrimage: living the presence and slowness during pilgrimage implies incorporating its traditional and historical condition, precisely as an alternative to the accelerated pace of everyday contemporary life; and reliving a pilgrimage right after it happened might be done through evoking an older past, as a nostalgia for a particularly authentic and intense, imagined past. Moreover, amid the affective publics of pilgrimage, different emotions and norms are contained and contested through and about digital media (Döveling et al, 2018; Ural, 2023). As much as there is mobilization and resonance, *communitas* and collective feeling (Berlant, 2008), there are also disputes and contestations regarding the position of this practice in collective social life (Couldry, 2003), which are more or less prevalent at times (for example, through memes). While post-secular forms of pilgrimage such as Santiago and WYD elicit wider discussion about how the 'authentic' pilgrimage should look, more traditionally religious pilgrimage to Fátima finds in social media new forms to keep its cultural relevance.

As they play out on social media and digital media more widely, these affective temporalities of pilgrimage cut through and recreate the linear and high-speed society (Howard, 2012; Wajcman, 2015). Our analysis decentred the discourses of disconnection from the notion of presentism and embraced the diffusiveness, complexity and ambiguity of digital atmospheres of pilgrimage. It acknowledges the tensions between compulsions to take pictures, post or share, and mandates to engage mindfully with the more solitary pilgrimage of Santiago, the religious experience of Fátima, or the sociable gathering of WYD, as constitutive of contemporary pilgrimage. Grounded in our investigation of both the multimodality of digital media platforms and services, and how pilgrims engage with them and negotiate norms surrounding them, our proposed overview of the affective rhythms of digital media use on pilgrimage practices can be explored in relation to other social events and gatherings.

Notes

[1] It is for this reason that the conceptualization of pilgrimage has been extended in media studies to look at fan and media tourism and pilgrimage – see Couldry (2007) or Williams (2017).

[2] https://passo-a-rezar.net/

References

Beckstead, Z. (2021) 'On the way: pilgrimage and liminal experiences', in Wagoner, B. and Zittoun, T. (eds) *Experience on the Edge: Theorizing Liminality*. Springer, pp 85–105.

Berlant, L. (2008) *The Female Complaint: The Unfinished Business of Sentimentality in American Culture*. Duke University Press.

Bogacz-Wojtanowska, E., Góral, A. and Jałocha, B. (2019) '"Catholic Coachella", "Papal Rock Concert"? Case study of the World Youth Day in Cracow as an example of a successful religious project', *Journal of Intercultural Management*, 11(2), pp 47–71.

Bouvier, G. and Rasmussen, J. (2022) *Qualitative Research Using Social Media*. Routledge.

Buitelaar, M. (2023) 'Narrativizing a sensational journey: pilgrimage to Mecca', in Buitelaar, M. and van Leeuwen, R. (eds) *Narrating the Pilgrimage to Mecca*. Brill, pp 1–42.

Caidi, N. (2023) 'Curating post-Hajj experiences of North American pilgrims: information practices as community-building rituals', in *Narrating the Pilgrimage to Mecca*. Brill, pp 369–390. Available at: https://library.oapen.org/bitstream/handle/20.500.12657/76308/1/9789004513174.pdf#page=386

Caidi, N., Beazley, S. and Marquez, L. C. (2018) 'Holy selfies: performing pilgrimage in the age of social media', *The International Journal of Information, Diversity, & Inclusion (IJIDI)*, 2(1/2). https://doi.org/10.33137/ijidi.v2i1/2.32209

Caminhos da Fé (2021) Caminhos de Fátima. *Caminhos da Fé* [online]. Available from: https://www.pathsoffaith.com/pt-pt/ways/caminhos-de-fatima

Coleman, R. (2018) 'Social media and the materialisation of the affective Present', in Sampson, T. D., Ellis, D. and Maddison, S. (eds) *Affect and Social Media*. Rowman & Littlefield, pp 67–75.

Collins-Kreiner, N. (2010) 'Researching pilgrimage: continuity and transformations', *Annals of Tourism Research*, 37(2), pp 440–456.

Conner, V. (2023) 'Fever dreams and the future of nostalgia on TikTok', in *AoIR Selected Papers of Internet Research*. https://doi.org/10.5210/spir.v2023i0.13408

Couldry, N. (2003) *Media Rituals: A Critical Approach*. Routledge.

Couldry, N. (2007) 'Pilgrimage in mediaspace: continuities and transformations', *Etnofoor*, 20(1), pp 63–74.

Couldry, N. (2012) *Media, Society, World: Social Theory and Digital Media Practice*. Polity.

Dickinson, J. E., Hibbert, J. F. and Filimonau, V. (2016) 'Mobile technology and the tourist experience: (dis)connection at the campsite', *Tourism Management*, 57, pp 193–201.

Döveling, K., Harju, A. A. and Sommer, D. (2018) 'From mediatized emotion to digital affect cultures: new technologies and global flows of emotion', *Social Media + Society*, 4(1). https://doi.org/10.1177/2056305117743141

Durkheim, E. (2001 [1912]) *The Elementary Forms of Religious Life*. Oxford University Press.

Edensor, T. (2016) 'Contemporary pilgrimage: journeys in time and space', in Maddrell, A., Terry, A., and Gale, T. (eds) *Sacred Mobilities: Journeys of Belief and Belonging*. Routledge, pp 201–210.

Egan, K. (2010) 'Walking back to happiness? Modern pilgrimage and the expression of suffering on Spain's Camino de Santiago', *Journeys*, 11(1), pp 107–132.

Ekelund, R. (2022) 'Fascination, nostalgia, and knowledge desire in digital memory culture: emotions and mood work in retrospective Facebook groups', *Memory Studies*, 15(5), pp 1248–1262.

Ekmekcioglu, C., Chandra, P. and Ahmed, S. I. (2023) 'A matter of time: anticipation work and digital temporalities in refugee humanitarian assistance in Turkey', *Proceedings of the ACM on Human-Computer Interaction*, 7(CSCW1), pp 1–36.

Firth, R. M., Rintel, S. and Sellen, A. (2020) 'Everyday time travel: temporal mobility and multitemporality with smartphones', in Kaun, A., Pentzold, C., and Lohmeier, C. (eds) *Making Time for Digital Lives: Beyond Chronotopia*. Rowman & Littlefield, pp 103–116.

franzke, aline shakti, Bechmann, A., Zimmer, M., Ess, C. M. and Association of Internet Researchers (2020) *Internet Research: Ethical Guidelines 3.0*. Association of Internet Researchers. Available at: https://aoir.org/reports/ethics3.pdf

Fundação JMJ Lisboa (2023) Festival da Juventude. *JMJ Lisboa 2023* [online]. Available from: https://www.lisboa2023.org/pt/festival-da-juventude

Germann Molz, J. and Buda, D.-M. (2022) 'Attuning to affect and emotion in tourism studies', *Tourism Geographies*, 24(2–3), 187–197.

Gonzalez, L. T. V., Mariz, C. L. and Zahra, A. (2019) 'World Youth Day: contemporaneous pilgrimage and hospitality', *Annals of Tourism Research*, 76, pp 80–90.

Heelas, P. and Woolhead, L. (2005) *The Spiritual Revolution: Why Religion is Giving Way to Spirituality*. Blackwell Publishing.

Hitchen, E. (2021) 'The affective life of austerity: uncanny atmospheres and paranoid temporalities', *Social & Cultural Geography*, 22(3), pp 295–318.

Howard, C. (2012) 'Speeding up and slowing down: pilgrimage and slow travel through time', in Fullagar, S., Markwell, K., and Wilson, E. (eds) *Slow Tourism*. Multilingual Matters, pp 11–24.

Hussain, T. and Wang, D. (2024) 'Social media and the spiritual journey: the place of digital technology in enriching the experience', *Religions*, 15(5), p 616.

Jansson, A. (2018) 'Rethinking post-tourism in the age of social media', *Annals of Tourism Research*, 69, pp 101–110.

Jansson, A. (2024) 'The cultured traveller: three theses on cultural capital and the taste for disconnection in tourism', in *The Digital Backlash and the Paradoxes of Disconnection*. Nordicom, University of Gothenburg, pp 345–362.

Jordheim, H. and Ytreberg, E. (2021) 'After supersynchronisation: how media synchronise the social', *Time & Society*, 30(3), pp 402–422.

Jorge, A. (2023) 'Pilgrimage to Fátima and Santiago after COVID: dis/connection in the post-digital age', *Mobile Media & Communication*, 11(3), pp 549–565.

Koivunen, A., Nikunen, K., Hokkanen, J., Jaaksi, V., Lehtinen, V., Soronen, A., Talvitie-Lamberg, K. and Valtonen, S. (2024) 'Anticipation as platform power: the temporal structuring of digital everyday life', *Television & New Media*, 25(2), pp 115–132.

Lefebvre, H. (2004 [1972]) *Rhythmanalysis: Space, Time and Everyday Life*. Continuum.

Lupinacci, L. (2024) 'Phenomenal algorhythms: the sensorial orchestration of "real-time" in the social media manifold', *New Media & Society*, 26(7), pp 4078–4098.

Lusa (2023) A Jornada Mundial da Juventude em números. *Público* [online], 31 July. Available from: https://www.publico.pt/2023/07/31/sociedade/noticia/jornada-mundial-juventude-numeros-2058624

Maddrell, A. and della Dora, V. (2013) 'Crossing surfaces in search of the holy: landscape and liminality in contemporary Christian pilgrimage', *Environment and Planning A: Economy and Space*, 45(5), pp 1105–1126.

Mascheroni, G. and Vincent, J. (2016) 'Perpetual contact as a communicative affordance: opportunities, constraints, and emotions', *Mobile Media & Communication*, 4(3), pp 310–326.

Miller, V. (2015) 'Resonance as a social phenomenon', *Sociological Research Online*, 20(2), pp 58–70.

Nehring, D., Alvarado, E., Hendriks, E. C. and Kerrigan, D. (2016) *Transnational Popular Psychology and the Global Self-Help Industry*. Palgrave Macmillan UK.

Nikunen, K. (2023) 'Affective temporalities of digital hate cultures', in Lünenborg, M. and Röttger-Rössler, B. (eds) *Affective Formation of Publics*. Routledge, pp 173–190.

Nilsson, M. (2018) 'Wanderers in the shadow of the sacred myth: pilgrims in the 21st century', *Social & Cultural Geography*, 19(1), pp 21–38.

Nilsson, M. and Tesfahuney, M. (2018) 'The post-secular tourist: re-thinking pilgrimage tourism', *Tourist Studies*, 18(2), pp 159–176.

Papacharissi, Z. (2014) *Affective Publics: Sentiment, Technology, and Politics*. Oxford University Press.

Pilgrim's Welcome Office (2025) Información Estadística Oficina Peregrino para 2025 hasta Abril. *Oficina da Acogida al Peregrino* [online], April. Available from: https://oficinadelperegrino.com/en/statistics-2

Pink, S. and Hjorth, L. (2014) 'The digital wayfarer: reconceptualizing camera phone practices in an age of locative media', in Goggin, G. and Hjorth, L. (eds) *The Routledge Companion to Mobile Media*. Routledge, pp 488–498.

Polus, R. and Carr, N. (2021) 'The role of communication technologies in restructuring pilgrimage journeys', *International Journal of Religious Tourism and Pilgrimage*, 9(5). https://doi.org/10.21427/MG4M-E412

Rosenberg, H. (2019) 'The "flashpacker" and the "unplugger": cell phone (dis)connection and the backpacking experience', *Mobile Media & Communication*, 7(1), pp 111–130.

Sanctuary of Fátima (2024) Santuário de Fátima acolheu 6,8 milhões de peregrinos nas celebrações em 2023. *Santuário de Fátima*, [online] 8 February. Available from: https://www.fatima.pt/pt/news/santuario-de-fatima-acolheu-68-milhoes-de-peregrinos-nas-celebracoes-em-2023

Santos, L. J. (2024) Quase meio milhão de pessoas nos Caminhos de Santiago. Portugal em alta. *Público* [online], 11 January. Available from: https://www.publico.pt/2024/01/11/fugas/noticia/quase-meio-milhao-pessoas-caminhos-santiago-portugal-alta-2076494

Schwarzenegger, C. and Lohmeier, C. (2021) 'Creating opportunities for temporary disconnection: how tourism professionals provide alternatives to being permanently online', *Convergence*, 27(6), pp 1631–1647.

Sebald, G. (2020) ' "Loading, please wait" – temporality and (bodily) presence in mobile digital communication', *Time & Society*, 29(4), pp 990–1008.

Serazio, M. (2013) 'The elementary forms of sports fandom: a Durkheimian exploration of team myths, kinship, and totemic rituals', *Communication & Sport*, 1(4), pp 303–325.

Sharma, N. and Timothy, D. J. (2023) 'Endurance rituals, performativity and religious tourism', *Annals of Tourism Research*, 100. https://doi.org/10.1016/j.annals.2023.103552

Sharma, S. (2014) *In the Meantime: Temporality and Cultural Politics*. Duke University Press.

Slavin, S. (2003) 'Walking as spiritual practice: the Pilgrimage to Santiago de Compostela', *Body & Society*, 9(3), pp 1–18.

Syvertsen, T. and Enli, G. (2020) 'Digital detox: media resistance and the promise of authenticity', *Convergence*, 26(5–6), pp 1269–1283.

Tosoni, S. and Turrini, V. (2018) 'Controlled disconnections: a practice-centred approach to media activities in women's solo travelling', in Peja, L., Carpentier, N., Colombo, F., Murru, M. F., Tosoni, S., Kilborn, R., Kramp, L., Kunelius, R., McNicholas, A., Nieminen, H. and Pruulmann-Vengerfeldt, P. (eds) *Current Perspectives on Communication and Media Research*. edition lumière, pp 283–302.

Turner, V.W. and Turner, E.L.B. (1978) *Image and Pilgrimage in Christian Culture: Anthropological Perspectives*. Columbia University Press.

Ural, H. (2023) 'Rethinking affective publics as media rituals through temporality, performativity and liminality', *Media, Culture & Society*, 45(5), pp 1036–1049.

Vilaça, H. (2008) 'Recomposições dos rituais contemporâneos: a peregrinação', *Sociologia: Revista da Faculdade de Letras da Universidade do Porto*, XVII–XVIII, pp 55–67.

Vilaça, H. (2010) 'Pilgrims and pilgrimages Fatima, Santiago De Compostela and Taizé', *Nordic Journal of Religion and Society*, 23(2), pp 137–155.

Wajcman, J. (2015) *Pressed for Time: The Acceleration of Life in Digital Capitalism*. University of Chicago Press.

Weltevrede, E., Helmond, A. and Gerlitz, C. (2014) 'The politics of real-time: a device perspective on social media platforms and search engines', *Theory, Culture & Society*, 31(6), pp 125–150.

Williams, R. (2017) 'Fan tourism and pilgrimage', in Click, M.A. and Scott, S. (eds) *The Routledge Companion to Media Fandom*, Routledge, pp 98–106. https://www.taylorfrancis.com/chapters/edit/10.4324/9781315637518-13/fan-tourism-pilgrimage-rebecca-williams

Wright-Ríos, E. and Martínez-Don, C.G. (2024) 'Posting the journey to Juquila: pilgrimage, digital devotion, and social media in Mexico', *Latin American Research Review*, pp 1–20. https://doi.org/10.1017/lar.2024.8

FOUR

Affective Intensities of Dis/connection in Mourning

Ionara Silva, Ana Jorge and Filipa Neto

Introduction

Grieving is shaped by personal factors as well as social relationships and cultural contexts, significantly influencing how people relate to and deal with experiences of loss (Hjorth and Cumiskey, 2018; Jacobsen, 2021; Pasquali et al, 2022). As online existence becomes permanent and the 'dead live on' (Burden and Savin-Baden, 2019; Lagerkvist, 2019), so does mourning (Sisto, 2020; Stokes, 2021). Bereaved people turn to social media to renegotiate the meaning of their loss over time (Kasket, 2020, 2023; Sumiala, 2022) as they attempt not so much to maintain 'continuous bonds' with their lost ones (Klass et al, 1996) but to keep hold of the deceased (Christensen and Sandvik, 2016). Yet, in some online spaces, timing or circumstances, sharing grief can be considered disproportionate or even inappropriate (Kneese, 2023; O'Connor, 2024), while in others where it is a communal experience it may be deemed more appropriate (Cumiskey and Hjorth, 2017; Hallam, 2018).

Public representations of death and grief through social media (for example, Gibbs et al, 2015 on Instagram; Eriksson Krutrök, 2021 on TikTok) may contribute to promote the visibility of a 'taboo' topic. At the opposite end of the spectrum, some scholarship emphasizes 'the risks [social media] entail, such as the erosion of privacy, the loss of control over the grieving and memorialization of loved ones and the inauthentic (over)sharing of emotion' (Giaxoglou, 2022, p 2). Our framework of affective atmospheres aligns with perspectives that highlight how both positive and negative aspects coexist in online mourning (Wagner, 2018), and new forms of discomfort and conflict may also arise from using social media and digital media more generally (Perluxo and Francisco, 2018). Social media platforms are not seen as monoliths, but rather as complex environments where different atmospheres can occur – affected by and affecting the different emotions their users go through.

The particular vulnerability of the mourning process can help illuminate the psychological, existential and relational ambiguities of disconnection, and therefore extend its conceptualization. On the one hand, as noted in the Introduction of this book, disconnection scholarship has been dominated by studies on individual and voluntary disconnection, associated with self-care and wellness. Emergent perspectives on disconnection by people with mental health problems or with special needs (Saukko et al, 2024; Beattie, 2025) are problematizing the notion of voluntary and calculated disconnection – and mourning can, we believe, add to this problematization. On the other hand, a relational perspective on disconnection means seeing not just what media people are disconnecting from, but *who* they are disconnecting – or indeed disconnected – from. Here, we draw on Markham's (2021) study and Šiša's work on ghosting (2024) to understand why *not communicating* with others, or others not communicating with us, may be disturbing.

This chapter thus focuses on how the affective intensities experienced during bereavement permeate connecting and disconnecting from digital media. From an early analysis of our data, it became salient that the complex emotional experiences throughout bereavement have implications for how the bereaved perceive and use media in general, and digital media especially.

We approach grief as a major life disruption that alters digital practices and routines, permeating longer-term, everyday social media uses (Döveling, 2017). Therefore, we look at the reconfiguration of digital media engagement by the bereaved through an ecological perspective (Treré, 2021). To unpack the affective intensities behind this reconfiguration, we explore the ecologies of dis/connection in mourning by looking at affective narratives (Stage, 2017; Kappelhoff and Lehmann, 2019; Giaxoglou, 2022) with existential and relational dimensions, focusing on: the bereaved and their *personal feelings* (ineffable grief); *the relationship with the deceased* (living on (memories)); and *engaging with other mourners* (communal support). These dimensions are heuristic and can therefore overlap. They do not mirror but can be seen in parallel with psychological models of mourning that identify a three-phase process: after a stage of numbness and loss, a period of intense pain follows, and acceptance and reorganization can occur (Perluxo and Francisco, 2018, pp 85–86). However, our analysis highlights how bereaved people's practices of dis/connecting to digital media, at each affective narrative, are riddled with ambiguities, tensions and existential struggles; they reflect conflicting desires to distance themselves from emotional overload *as well as* connection and presence with the deceased, other mourners and their social circles.

Methods and materials

In countering the universalization of the experience of bereavement in literature (O'Connor, 2024), we acknowledge the cultural, political and economic factors that shape those experiences (Maddrell, 2020) and focus on the Portuguese-speaking territory of Portugal and Brazil. Moreover, we included 'disenfranchised mourning' such as mourning due to miscarriage and suicide (Christensen and Sandvik, 2016; Perluxo and Francisco, 2018); and extended our attention to immigrants or racialized people (Doka, 2008; Turner and Stauffer, 2023).

Similar to other studies in our research project, as presented in the Introduction, we used qualitative methods in a two-part study.

(1) A multimodal analysis was conducted on nine bereavement Facebook Portuguese-language groups. Navon and Noy (2023a) emphasize that Facebook pages, groups and profiles are *sub-platforms* that allow for different social dynamics. We included groups with different sizes, and public/private status, for two months. The Facebook profile of the first author was preferred over a blank profile. We conducted an initial exploration with keywords in Portuguese such as *grief, pain, I lost my son, heart in mourning, grieving mothers, grieving parents, COVID-19 mourning*. After identifying relevant Facebook private groups, the main author responded to the moderators' questionnaire and disclosed the intention for non-participant observation. Only one of the groups did not allow access.

Three groups brought together people who lost someone during the COVID-19 pandemic; data were collected between November and December 2021, when there were still sanitary restrictions in place in Portugal, and close to Christmas. The other six Facebook groups were dedicated to parental bereavement, and data was collected between January and February for three of them, and between June and July 2023 for the other three. The material from private groups was scraped manually and organized in Excel, with posts saved separately as screenshots or videos. The final corpus amounted to 940 posts. Group size ranged from 3,800 members to 31,000 at the time of data collection (average of 17,700 members); four groups were private and five were public. From the description or the moderators' presentation, we could detect that five groups were based in Brazil, three in Portugal, and one had members from both countries. Groups were anonymized and attributed a letter (A to J); in presenting results, we include the characteristics of each group in the description. Generally, groups dedicated to mourning under COVID-19 restrictions brought together mostly young to adult participants, and, except for group A, the cause of death was not specifically COVID-19 infections;

as for groups dedicated to communities of people who lost a child, they were mostly populated with female participants (as other studies on different types of loss have found, for example, Döveling, 2017) who had lost children of any age, and by different causes of death (health, accident or crime).

Social media research followed the Association of Internet Researchers' (AoIR) *Ethical Guidelines* (franzke et al, 2020). After seeking consent from the group moderator, all material, including public Facebook groups, was anonymized, and identifiable attributes were removed; it was saved on hard drives only accessible to the team of authors. When including exemplary text, URLs are not included, names are anonymized and translation prevents identification; we refrain from using exemplary images, and instead use descriptions when necessary.

(2) We conducted an interview study between December 2023 and February 2024, using a semi-structured format, with (a) five moderators of bereavement support groups in Portugal, organized by formal associations and a palliative care company. These took place on Zoom or face-to-face for a total of 3 hours and 23 minutes, and were summarized in Word documents. We also interviewed (b) 10 people in mourning, aged 18 or above, living in Portugal and native or fluent Portuguese speakers, who had lost someone between 2020 and early 2023. We shared these criteria with the moderators, who negotiated access to the group of participants; we also advertised the recruitment on the researchers' social media profiles. We invited 19 people for interviews, trying to balance the gender ratio, but were only able to recruit one man for the study, which stands as a limitation and evidence of a more invisible mourning process among men (Döveling, 2017). The participants' characteristics are presented in Table 4.1. The interviews took place at times of convenience for the participants, and nine of them chose to do the interview over Zoom. In the case where the interview took place face-to-face, the interviewee chose to do it in their workplace after work hours, with some noise and distractions. The subject of loss led the interviewee to cry, and after some time, the researcher

Table 4.1: Characteristics of the sample of bereaved interviewees

Name/pseudonym	Age	Nationality	Occupation	Most recent loss	Cause of death
Bruna	40	Brazilian	Bus driver	Sibling	Suicide
Joana	23	Brazilian	Influencer	Mother	Illness
Rina	39	Italian	Professor	Mother	Illness
Maria	36	Portuguese	Manager	Mother	Illness
Liliana	41	Portuguese	Manager	Daughter	Miscarriage
Raquel	31	Portuguese	Psychologist	Sibling	Illness
Manuel	40	Portuguese	Cultural producer	Mother	Suicide
Isabel	48	Portuguese	Doctor	Mother	Illness
Carla	65	Portuguese	Biologist	Mother	Natural cause
Bárbara	26	Portuguese	Nail designer	Daughter	Miscarriage

interrupted the interview to protect both the participant and herself. In Zoom interviews, the participants became emotional on several occasions; when this happened, the researcher deployed active listening, empathy and gave the participants time before returning to the interview. The interviews ranged from 27 to 65 minutes (average of 44 minutes). The audio recording was transcribed verbatim, and additional notes were taken regarding the settings, emotions and expressions during the interviews.

Interviews were conducted by the first author, who disclosed her position as a mourner to the potential participants. Given the participants' vulnerability, besides an informed consent form, with an option for termination at any time and for participation with a pseudonym that were common to other research in the project, a dedicated Contact Guide for emergency psychological counselling support was handed out, with contacts of institutions that provide free and emergency psychological support. After the interview, a follow-up email was sent to the interviewees to thank them and allow for prolonged contact with the researcher. Furthermore, anticipating the possibility of 'moral and emotional unease' for the researcher(s) themselves (Johnson and Clarke, 2003), the team of authors had iterative rounds of discussion after the interviews took place, and their transcription, performed by the third author. Interviews were saved on a hard drive accessible only to the authors and will be deleted five years after the project ends. The codebook was used for both social media analysis and interviews; it focused on media and platforms as environments; forms of digital interaction; the tensions and struggles of being digitally present during grief; affective aspects related to grief; tactics of disconnection; support groups; and digital media use during grief.

Ineffable grief ♥

Our interviewees identified Facebook feed as the space to mainstream the funeral details immediately after receiving news of the death, and to receive condolences on the bereaved person's profile. However, there was a need to avoid an overflow of interactions from friends and family, such as instant messages, posts and comments. For example, Rina mentioned finding it overwhelming to deal with other people's feelings while dealing with her own pain. The mourners' interruption of social media presence is often accompanied by sadness, not knowing how to talk about their emotions, or even not having anything to talk about besides grief, which could, in turn, overwhelm others. Deciding what to share on social media about mourning is a complex task given the pervasive social norms about what is private and public, and what should be kept intimate versus shared publicly (Hjorth and Cumiskey, 2018; Pasquali et al, 2022).

While there is no pre-fixed temporality for mourning, the bereaved reported temporarily disconnecting from social media during the first months after losing someone. They adopted disconnection tactics to avoid physical and emotional overload and fatigue caused by having to deal with after-death issues, such as funerals, heirlooms, banks, state-related bureaucracies, as well as digital objects and legacies. Many of our participants reported avoiding interactions, both online and offline, because the pain of losing a loved one is 'unimaginable' and 'incomparable'. Talking about the loss, especially during the first months, and expressing how one feels might be extremely hard (Perluxo and Francisco, 2018). The seriousness of the grief, as well as the feeling that others are unable to understand the pain and ambivalence of feelings they are experiencing, often dictates ineffability. Joana explains how she lost her mother in her home country, Brazil, and how she could only spend a week with her family; two years after, she comments how 'if I talk about it, I still cry'. She has told very few people about her mother's passing, as talking about it still made her feel uncomfortable. In the first few months of mourning, the pandemic exacerbated her feelings of isolation and amplified the emotional difficulties associated with mourning.

While Carla and Isabel tell us that they were already inactive on their social media profiles before their loss, and remained that way, to Joana and Liliana, being silent about their loss stood against their previous actions, or out of their everyday online activity, as they feel it is misplaced on social media, because such issues are intimate or hard even to articulate. Again, Joana, who works as an influencer and lost her mother, says she was unable to publish anything related to mourning on her Instagram account at the time, and ever since. She says other people advised her to 'take advantage of that [the loss] to post it, [because] other people would like to see it', but she cannot, 'because it hurts too much. I don't even know. The pain of absence, of knowing that the person will not be here anymore'. She recalls how, a few weeks after her mother's death, she was invited to go to a big music festival, and how she was joking around and interviewing people for online content, but that she felt 'mortified'. The Portuguese expression, etymologically related to death, conveys how disheartened and conflicted she felt to be visible while hiding her grief.

For Liliana, who had posted pictures of her pregnancy and the ultrasound scan with the baby's sex, against her family's advice, the feeling was different after she had a miscarriage: 'It was … it was a year in which I did not post anything because there was nothing I wanted to say to the world.' She felt that if she were to post about anything, it would be 'false'. So, she preferred to engage with other online content, namely TV series.

For some participants, mere scrolling on social media platforms became 'toxic'. They draw on a vulgarized term to talk about social media and describe their discomfort and feeling off the general tone of the platforms, or the tone of their own profiles. Isabel emphasized that at one point she accessed certain types of content that aroused 'toxic feelings' in her. Bárbara felt that her relationship with social media was doing her 'harm', especially in the early days of her bereavement. Joana confessed that she 'did not have the strength' to talk about death on her profiles, which she considered to be 'cheerful' and entertaining. In Rina's case, she was conflicted and confused about sharing her deep pain among trivialities such as 'complaints about broken nails', but she also wanted to respect the space for such everyday affective expressions.

While social media posts do not give us information about the time or stage of grief any person is at, the expression of personal feelings and affective states was frequent on Facebook groups (Navon and Noy, 2023b). The bereaved expressed different emotions such as anger, sadness, joy, gratitude, indignation or discomfort. Often, this is preceded by expressions such as 'I feel', 'I lost', or 'I miss'. One example from a Portuguese-based, private group devoted to people mourning the loss of their child reads, 'Do you know that sadness that won't go away, won't go away and won't get better? That's what I'm feeling now.' As we will explore below, turning to closed groups might be a strategy to evade judgement from non-mourners and seek and reciprocate support with other mourners.

While our interviewees refrained from posting or even accessing social media feeds for self-protection, they described investing heavily in consuming and engaging with information about death and grief, especially after a few months of the loss. Reading, listening and watching other people's – current or past – experiences of losing a loved one ultimately made them feel that grief is a communal experience. The bereaved move between Google searches as their first source of content that can help them understand the grieving process. Preferred formats included podcasts, YouTube videos, blogs, profiles and pages on Instagram and Facebook, in the areas of personal development and self-help, as well as films and series. Bárbara tells us that some of this content was 'tools provided by the [support] group', in the form of interviews with doctors, therapists or coaches on YouTube videos; she also watched Instagram Lives by accounts (she perceived as) experts. Other interviewees mentioned they followed profiles of bereavement support groups, such as associations or informal groups, and companies providing palliative care services. To Bruna, reading about grief on social media helped her understand that what she was feeling was not wrong. Bárbara, however, felt tired of these contents after a while, 'Because I saw many people who didn't understand the experience [grief] talking about it. And I thought this could be misleading for some people, meaning that instead of helping, it would have the opposite effect.' Acknowledging the credibility of others to *speak* about grief is paramount here.

This commercialized and less personalized advice on content accessed online was a resort for some of our participants, especially migrants and women with lower education levels, who did not have (easy) access to psychological support and other services. Bruna had three counselling sessions offered by her employer (but ultimately gave up because she did not feel that her bereavement experience was understood); and Joana did not have any professional support, which is rarely offered by the public health system in Portugal.

Other participants told us they adopted therapeutic practices, psychological counselling, dance classes, yoga and meditation. For some, mourning during the pandemic restrictions meant these activities were conducted online; for example, Rina's dance classes migrated to a specific WhatsApp group, and Carla held her meditation sessions on Telegram. As the restrictions eased off, the interviewees started carrying out face-to-face activities. Isabel, for example, realized the long-held wish of walking the Camino de Santiago/Way of Saint James alone, and she adds that she 'was not on social media on the way', like many participants in Chapter 3 report. Isabel tells us she received the advice to go on a pilgrimage from a nun, and that it was her way of dealing with her own thoughts, constantly referring to her grief as 'indescribable'.

Living on (memories)

Digital devices such as smartphones and social media become repositories of memories and legacies of deceased loved ones

(Stokes, 2021). However, there is also a conflicted affect in the relationship the bereaved establish with the digital presence of the deceased (Lagerkvist, 2019): on the one hand, they avoid interaction or contact with the profiles of the deceased, most of which are still active or have been transformed in memorials; on the other, they fear losing access to memories and experiencing definitive loss (Stokes, 2021). Bruna, for example, explains how losing posts might mean losing memories. Maria tells us that her mother's Facebook profile is 'a database of ... photos, of memories of her', so she does not want to remove the profile.

Despite looking at older or more recent posts, our participants tell us that, in the first months of mourning, accessing social media causes them unease because it pushes them into grief. Among our interviewees, during the first months of mourning, accounts of the deceased remained suspended, and interviewees did not necessarily delete accounts or remove profiles. Although our interviewees said they did visit the deceased's profile to look at old posts and memories of the loved one (be it through the profile, birthday notifications, archives and/or memories selected by Facebook's 'On This Day'), doing this caused them discomfort.

While for Joana, notifications of memories with her deceased mother acted as happy reminders, for others, notifications could act as triggers, especially in the case of bereavement for siblings and suicide, when death is unexpected and felt unnatural, too soon. The most resistant participants were Raquel and Carla; the latter expressed her indignation at unsolicited notifications from Google Maps or Photos: 'I didn't ask for it, I mean, I was not even happy to receive that photograph. In fact, it bothered me. I felt "But why?", that's getting into my intimacy.' For Raquel, who lost her sister with whom she used to interact on social media often, notifications are not 'a bad thing', but they are a Sisyphean, eternal repetition, 'just constantly living, experiencing the same thing', as they can no longer interact.

As a result, some interviewees avoided notifications as a measure of self-protection. More intermediate, and affective, responses were illustrated by Bárbara, who was positive about notifications but still decided to opt out of them.

Kaun and Stiernstedt (2014) observed that Facebook is 'an atmosphere and an interface of rapid change and forgetfulness, rather than of remembrance and preservation' (p 1161), but it does afford possibilities for 'remembrance, archiving, preservation, and stasis' (p 1162). For mourners, these possibilities were important to memorialize the deceased on social media and keep their memory alive, not to let their life story be forgotten by their family and network of friends, a form of comfort in the face of the complexity of mourning (Kasket, 2023). The bereaved find validation for their feelings in interactions with other users and reaffirm a positive memory of the deceased. Even if the relationship was not close, or in the face of geographical distance, the bereaved perform emotions that they did not necessarily do in their online narratives before the death of their loved one. However, this can also act as a pressure the bereaved put on themselves, as a need to sustain the deceased person's memory for their network of family and friends. Maria, for example, mentions posting on her mother's birthday as a way to express how her mother is still present in the family, seeing that not everyone remembers when her birthday was.

Our interviewees demonstrated online memorialization practices during specific periods of mourning. The most common is posting messages or tributes on specific and/or commemorative dates; all interviewees mentioned posts on birthdays and death anniversaries to honour their deceased loved ones. Even if the bereaved temporarily suspended their online profiles, they return to remind themselves and everyone else that this person existed. They post on their own feed as a 'personal mourning log' or interact with the dead person's profile as an 'online mourning guestbook' (Navon and Noy, 2023a), and carefully curate photographs and text: they select the pictures that best represent, in their opinion, the person when they were alive. Raquel kept up the habit of celebrating her sister's birthday with a post on her (Raquel's) feed, as she always did when her sister was alive, as a strategy of not interrupting the bonds and rituals that can no longer be carried out, but which are not forgotten. In Maria's case, in addition to her mother's birthday messages on Facebook, she and her family got together to 'celebrate her life'; they went to the beach and made records as a strategy to keep her mother present in the family. Maria was an exceptional case in our sample who hired a doula to prepare her mother and the family for her mother's passing. She says that sharing the end of her mother's life on social media was a 'mission' she had because it was 'very beautiful and with much love'.

The absence of the deceased, like 'ghosting' in romantic relationships (Šiša, 2024) – again, etymologically under the umbrella of death – creates an abrupt and unilateral cease of communication. As this absence translates into silence, with no 'echoes' or responses (using Markham's 'social echolocation' theory (2021)), mourners may incline to question their own existence and be led to emotional states of anxiety and disorientation. Even if there is no response, some public demonstrations include a direct dialogue with the deceased, as if they were listening or reading the message. This addressivity strategy aims to prolong the bonds (Klass et al, 1996; Kasket, 2012). In a private Facebook group, a mother writes, 'Today I thought of you and missed you and my heart was filled with sadness, and it made tears overflow through my eyes. But there's nothing new about it, because I thought about you yesterday and the day before and every day since you left' (group J, lost a child, Portugal). In another private group, a participant writes: 'Today it would be my mother's birthday, I'd be baking her cake as usual! ... I miss you forever mother, I wanted to hug you so much ❤[broken heart]' (group A, COVID-19, Brazil).

In the interviews, Manuel, who lost his mother to suicide, tells us he always posts messages on special dates, such as his mother's birthday, and leaves messages on her profile as if he were talking to her, about things he wishes he had said to her, but which were never spoken when she was alive. However, he says these posts have never included photographs of his mother. Moreover, when others address the deceased and express grief (Wagner, 2018), the mourners who feel they are the most entitled to grief may resent it or feel disturbed, or it may bring reassurance. Raquel explains how seeing people writing on her late sister's profile page as if she were alive 'bothers' her, but she admits that reading these messages brings her parents 'some peace of mind'.

While Manuel refrains from posting pictures of his mother, for Rina, posting pictures is crucial to her memorialization. Rina told us that her parents did not have online profiles, so all the pictures of them online were posted by her. Since her mum passed away, she 'started celebrating anniversaries of life and death, mine and theirs, with images of them', for instance, on her mother's birthday; these included photographs from when they were young and even before she was born. She says, 'It's a way of also … indirectly … to be a bit connected with … with my old folks.' She also finds it very supportive that her parents' close friends usually comment on these pictures or message her privately afterwards, especially since she has migrated from their home context. However, this practice raises concerns about the deceased's right to be forgotten, image rights and how they might wish to be portrayed (Fordyce et al, 2021).

Memorializing or addressing the deceased on a personal feed can, in turn, expose the mourner to judgement or lack of understanding from social contacts (Wagner, 2018), especially if there is a sense that they are *oversharing*, that is, that the publics of those expressions feel it is too intimate or prolonged (Kennedy, 2018). Thus, performing this narrative among other mourners may be safer, as the next section will explore. This helps to make sense of the many occurrences of this narrative across our Facebook groups material, where usually the mourner addresses the deceased to share their feelings on a special, or an ordinary, day: 'Today you would have turned 75 … and all I can do is miss you … how much

I miss you. I love you so much (…)' (public group, COVID-19, Brazil). On Facebook groups, many posts included original or edited pictures where the deceased appears, as proof of, and reinstating, the intimacy and joy of having been together. We also found many instances of re-shares – of Facebook Memories, posts of publicly available content – and GIF art, modified pictures with stickers and overlaid text, and e-postcards from repositories or saved by the users.

Memorialization appeared to be particularly uncomfortable for mourners in cases of miscarriage and suicide. The stigma and taboo intrinsically linked to religious, cultural and privacy issues complicate the bereavement, marked by ambiguities and restrictions. Miscarriage can be impactful and painful for the parents, especially when there is a diagnosis of a fetal anomaly that could affect their survival, limit their health and interfere with their development, as well as pose risks for the pregnant woman. Memorializing such a loss can help maintain a connection between parents, especially mothers, and the deceased fetus, and help with the affective aspects related to absence (Christensen and Sandvik, 2016).

However, as mourners of miscarriage seek not to let the memory of their babies disappear within the family and reconstruct their grief as valuable (Stroebe et al, 2010), they feel their grief is often neglected. The mothers we interviewed, Liliana and Bárbara, reported feelings of displacement, loneliness and incomprehension, bringing them back to memorialization practices based on ultrasound scans and videos recording heartbeats, which are at once 'religious' (Liliana) and painful to watch. Often, the family, and society as a whole, repress affective demonstrations through unconscious discourses that a new pregnancy will soften the effects of mourning, which the mothers vehemently reject as aggression.

Here, the absence of interaction, or lack of recognition, from others makes the bereaved feel disconnected (Markham, 2021) as it creates a gap that amplifies feelings of isolation that they experience. This is also the case for Bruna, who lost family members in Brazil while she was living in Portugal. She uses social media platforms to express her pain, which was met with contempt from other family members who attempted to minimize her pain, seeing as she was far away (O'Connor, 2024). Bruna felt disconnected, isolated and helpless even while, or because, she was using social media to express her grief.

Communal support

In tandem with avoidance and feeling sanctioned or disconnected for their public grievance, mourners in our interview and Facebook sub-studies use support groups and/ or online groups to meet and interact with other people in grief in search of emotional sharing, support and connection (Hartig and Viola, 2016). This section analyses the elements that make these online or offline spaces safe and how they form communities of affection, care and support. Mourners expect that people who experienced similar pain, longing, or sadness will be able to understand and respect them, and be empathetic and accepting (Perluxo and Francisco, 2018). This sense of belonging aligns with what Berlant (2008) describes as *communitas* – an affective intensity generated by the perception that others 'feel the same way'.

In online semi-public environments such as Facebook groups, grief communities function as safe spaces for emotional expression, as we have alluded to, where participants share vulnerabilities and feelings without fear of judgement, which can enact emotional validation and mutual support (Brubaker et al, 2013; Martinuzzo and Sangalli, 2019). The Facebook groups we analysed were often created and moderated by bereaved people. One moderator of a private group dedicated to people mourning the loss of their child writes: 'This group aims to help you grieve spiritually, physically, and emotionally. … We need support in grief ♥[broken heart].' To our interviewee Liliana, being involved in a support group was not viable in an earlier stage of grief; however, three years after her pregnancy, she acknowledges the change, saying how now she participates 'with the perspective of helping others rather than receiving help'.

In face-to-face grief support groups, the atmosphere is more contained, as the groups are smaller, with 8 to 15 participants, and moderation is more present. In these meetings, the bereaved can talk with people going through similar processes and feel heard and supported. Mourners feel safe to express feelings, cry, or remain in silence; other listeners sit and give them time. Bárbara explains how 'the most important thing for me was to feel that I was surrounded by a group of people who, because we have all been through the same thing, can understand the experience'. Support groups promoted by associations and a palliative care company seek to extend these supportive atmospheres through social media, offering resources and promoting online meetings or live events. Bárbara tells us that meeting face-to-face was more frequent, with few online meetings; but she felt engaged and supported through dedicated social media lives, for instance. Carla, however, participates online reluctantly, as she struggles to use the digital technology because she 'belongs to another generation', but feels she must, so as to avoid 'missing out'.

Reciprocity is key to this form of 'intimate publics' (Berlant, 2008; Gibbs et al, 2015; Kennedy, 2018). In the Facebook groups we analysed, we found numerous accounts of personal experiences, messages of support and prayers or songs directed at helping or comforting others. Through text, images, emojis, music or videos, these messages generate a shared consciousness where other bereaved people come together to comment on and influence how the bereaved process their experiences (Hjorth and Cumiskey, 2018, p 24). Some members trust these environments and their members to reveal images with identifiable elements, personal data, routines and beliefs. For example, in a public group based in Brazil, the moderator seeks to recruit other mothers who have lost a child to join the group by posting a photo of her and her late son, urging anyone to leave a phone number, or to contact her, leaving her number.

Moreover, this affective public (Papacharissi, 2014) is also maintained through phatic communication, in the form of greetings, thanks or reactions. In the context of the Facebook groups we studied, the most common reactions were thumbs up (👍), crying face (😢) and care (🤗). In this context, like can sometimes represent an acknowledgement of the post, as a bare form of support, and not be interpreted as an appreciation of the grief that is being expressed. This myriad of everyday, small interactions reinforces the symbolic connection between members and forms the basis for a transnational and transpersonal public sphere where strangers share emotions and experiences in a de-spatialized environment (Gibson and Talaie, 2018). Our study explored a Portuguese-speaking territory of grief, where mostly Brazilian and Portuguese members could be discerned. Indeed, grief needs to be communicated in a culturally situated way (Perluxo and Francisco, 2018), and language plays a fundamental role in this context. This weight of shared language and cultural context in the expression and connection around grief also helps to explain why Rina, an Italian woman we interviewed, found that she could better express her grief in a WhatsApp group in her mother tongue and with people who understood the context where she grew up and her parents died.

As mentioned, Facebook groups dedicated to communities of people who have lost a child were populated mainly by female members. The testimonies revolve around the loss of children, but also husbands and fathers. While this might, again, indicate gendered norms about grief, it reinforces the homogeneity and identity of these groups as spaces to express female or maternal pain and seek solidarity. Cultural norms, however, are discernible, as members from Brazil expressed more religious content, which can be connected to care cultures in the Global South, where 'the dead are seen as spiritually present and in need of ongoing care' (Walter, 2017, p 27). An example of this, in a public Facebook group, is: 'My strength comes from God, only Him can help us go through this great pain. Together we can help each other to go through the pain, count on me, mothers of angels 😇🙏[smiling face with halo, folded hands]' (group E, lost a child, Brazil). As part of Western Europe, Portugal exemplifies memory cultures, which 'emphasize remembrance and legacy, viewing the dead as gone but symbolically alive through memory' (Walter, 2017, p 27).

As described earlier in the chapter, psychology models of mourning identify a three-phase process: after a stage of numbness and loss, a period of intense pain follows, and acceptance and reorganization can occur (Perluxo and Francisco, 2018, pp 85–86). It is necessary, of course, to consider the type of death, as unexpected deaths due to accidents or sudden illness do not allow for preparation for the end, as prolonged illness or natural causes owing to age do (p 86). Suicide as an internal volition for death is also relatively more complex to process than external causes. For our interviewees, acceptance and reorganization occurred around 6 to 12 months after the loss. Participating in selected online and offline safe spaces is perceived by the bereaved as essential for reintegration into social life after the loss. While not completely abandoning these grief support spaces, our interviewees reported that, progressively, they return to other online environments, at the same time as they return to social offline activities.

Conclusion

Mourning is a complex field where affect, social norms and platforms converge. Digital media, particularly social media platforms, do not merely mediate grief but actively reshape how it is processed and shared (Kappelhoff and Lehmann, 2019). Our analysis highlighted how emotions evolve with the mourning process *and* the reconfiguration of connection and disconnection. The emotional turmoil of grief leads some to suspend social media use or modulate platforms, evading an overflow of interactions, painful memories from the deceased and 'cheerful' social media feeds. This detachment can be seen as a form of self-protection and self-care from atmospheres that feel 'toxic', harmful or inappropriate, or as sheer inability to express, communicate, find others or expose oneself to the risk of not being acknowledged in the deep pain one is experiencing.

By revealing the complexity and ambiguity of dis/connecting throughout mourning, this chapter contributes to overcome binary discussions over the intersection of death and media (Giaxoglou, 2022). While social media offers mourners spaces for remembering and honouring the dead that go beyond conventional funeral traditions, and affords the possibility to people who are geographically distant to grieve at a distance, social norms draw the boundaries to accept expressions of grief. Different members of the family and broader social circles may have different, indeed conflicting, views on how memories should be held, leading to tensions and disagreements. The lack of recognition of grief or felt sanctioning can generate an internal disconnect that aggravates the pain of loss.

Our interviews and multimodal analysis revealed how practices of digital dis/connection, as in everyday life, of people in grief are fraught with affective tensions between contradictory desires for connection and separation, presence and absence, remembering and letting go. Throughout, online interactions can serve as both a space of comfort and remembrance, and a confusing and challenging terrain where social expectations can generate anguish. Algorithmic mediation is rejected by some as it acts as triggering, unsolicited or just repetitious, and embraced by others as joyful memories. This chapter thus contributes to problematize voluntary and calculated accounts of disconnection, and attunes to views on disconnection out of distress, where disconnecting may actually aggravate such state.

Both our studies reached mostly female accounts and discourses. Women's experiences in mourning seem to extend their care work and reinforce traditional gender roles (Olson, 2022; Kneese, 2023) as they memorialize their lost ones, express emotions and connect with others. Not only does this labour require digital literacy and the emotional effort of coordinating different members of the family or mourners, but it is also relatively invisible and might result in an additional emotional toll. Moreover, our findings from a Portuguese-Brazilian sphere evidenced different memorialization cultures brought together under the same language. Care cultures in the Global South may represent additional pressure to memorialize the deceased on social media, and put migrants at special tension between different norms and their role as mourner at a distance.

Among our group of interviewees, it became apparent that taking up 'offline' practices such as yoga, retreats, therapies and therapeutic support groups was a marker of socioeconomic privilege, as the public health system in Portugal does not guarantee access to mental health care and psychological counselling. Moreover, those practices seemed to depend on a high level of literacy and information (or knowing where to look for help), and social norms related to emotional openness to process bereavement and experiencing (online and offline) retreat from interactions. The exceptional case of hiring a doula to support the person dying and their family, recounted by Maria, signalled economic and cultural distinction, which appeared to have provided extraordinary emotional support. In contrast, migrants, people from lower social classes or in precarious labour situations did not have easy access to therapeutic support. Our difficulty in recruiting male participants might indicate that gender roles might also push men away from private support for grief. The consumption of dedicated social media and streamed content on grief might also derive from literacy. Moreover, while a public debate from narrated personal experiences can generate both social and economic value (Stage, 2017) and be complemented by expert involvement, it is important to discuss the fact that it offers universalized, rather than individualized, advice and support.

References

Beattie, A. (2025) 'ADHD and digital disconnection: exploring inclusive and practical approaches', *Media, Culture & Society*, 47(4), pp 805–814. https://doi.org/10.1177/01634437251326482

Berlant, L. (2008) *The Female Complaint: The Unfinished Business of Sentimentality in American Culture*. Duke University Press.

Brubaker, J. R., Hayes, G. R. and Dourish, P. (2013) 'Beyond the grave: Facebook as a site for the expansion of death and mourning', *The Information Society*, 29(3), pp 152–163.

Burden, D. and Savin-Baden, M. (2019) *Virtual Humans: Today and Tomorrow*. 1st edn. Chapman and Hall/CRC.

Christensen, D. R. and Sandvik, K. (2016) 'Grief and everyday life: bereaved parents' negotiations of presence across media', in Thorhauge, A. M. and Valtysson, B. (eds) *The Media and the Mundane*. Nordicom, pp 105–118.

Cumiskey, K. M. and Hjorth, L. (2017) *Haunting Hands: Mobile Media Practices and Loss*. Oxford University Press.

Doka, K. J. (2008) 'Disenfranchised grief in historical and cultural perspective', in Stroebe, M. S., Hansson, R. O., Schut, H. and Stroebe, W. (eds) *Handbook of Bereavement Research and Practice: Advances in Theory and Intervention*. American Psychological Association, pp 223–240.

Döveling, K. (2017) 'Online emotion regulation in digitally mediated bereavement. Why age and kind of loss matter in grieving online', *Journal of Broadcasting & Electronic Media*, 61(1), pp 41–57.

Eriksson Krutrök, M. (2021) 'Algorithmic closeness in mourning: vernaculars of the hashtag #grief on TikTok', *Social Media + Society*, 7(3). https://doi.org/10.1177/20563051211042396

Fordyce, R., Nansen, B., Arnold, M., Kohn, T. and Gibbs, M. (2021) 'Automating digital afterlives', in Jansson, A. and Adams, P. C. (eds) *Disentangling: The Geographies of Digital Disconnection*. Oxford University Press, pp 115–136.

franzke, aline shakti, Bechmann, A., Zimmer, M., Ess, C. M. and Association of Internet Researchers (2020) *Internet Research: Ethical Guidelines 3.0*. Association of Internet Researchers. Available at: https://aoir.org/reports/ethics3.pdf

Giaxoglou, K. (2022) 'Affective positioning in hyper-mourning: sharers as tellers, co-tellers and witnesses', *Conjunctions*, 9(1), pp 1–12.

Gibbs, M., Meese, J., Arnold, M., Nansen, B. and Carter, M. (2015) '#Funeral and Instagram: death, social media, and platform vernacular', *Information, Communication & Society*, 18(3), pp 255–268.

Gibson, M. and Talaie, G. (2018) 'Archives of sadness: sharing bereavement and generating emotional exchange between strangers on YouTube', in Dobson, A. S., Robards, B., and Carah, N. (eds) *Digital Intimate Publics and Social Media*. Springer, pp 281–297.

Hallam, E. (2018) 'Death and digital media: an afterword', in Arnold, M., Gibbs, M., Kohn, T., Meese, J. and Nansen, B. (eds) *Death and Digital Media*. Routledge, pp 141–156.

Hartig, J. and Viola, J. (2016) 'Online grief support communities: therapeutic benefits of membership', *OMEGA – Journal of Death and Dying*, 73(1), pp 29–41.

Hjorth, L. and Cumiskey, K. M. (2018) 'Affective mobile spectres: understanding the lives of mobile media images of the dead', in Papacharissi, Z. (ed) *A Networked Self and Platforms, Stories, Connections*. Routledge, pp 111–124.

Jacobsen, M. H. (2021) *The Age of Spectacular Death*. Routledge.

Johnson, B. and Clarke, J. M. (2003) 'Collecting sensitive data: the impact on researchers', *Qualitative Health Research*, 13(3), pp 421–434.

Kappelhoff, H. and Lehmann, H. (2019) 'Poetics of affect', in Slaby, J. and Von Scheve, C. (eds) *Affective Societies: Key Concepts*. Routledge, pp 210–219.

Kasket, E. (2012) 'Continuing bonds in the age of social networking: Facebook as a modern-day medium', *Bereavement Care*, 31(2), pp 62–69.

Kasket, E. (2020) *All the Ghosts in the Machine: The Digital Afterlife of Your Personal Data*. Robinson.

Kasket, E. (2023) *Reboot: Reclaiming Your Life in a Tech-Obsessed World*. 1st edn. Elliott & Thompson.

Kaun, A. and Stiernstedt, F. (2014) 'Facebook time: technological and institutional affordances for media memories', *New Media & Society*, 16(7), pp 1154–1168.

Kennedy, J. (2018) 'Oversharing is the norm', in Dobson, A. S., Robards, B. and Carah, N. (eds) *Digital Intimate Publics and Social Media*. Springer, pp 265–280.

Klass, D., Nickman, S. L. and Silverman, P. R. (eds) (1996) *Continuing Bonds: New Understandings of Grief*. Routledge.

Kneese, T. (2023) *Death Glitch: How Techno-Solutionism Fails Us in this Life and Beyond*. Yale University Press.

Lagerkvist, A. (2019) *Digital Existence: Ontology, Ethics and Transcendence in Digital Culture*. Routledge.

Maddrell, A. (2020) 'Bereavement, grief, and consolation: emotional-affective geographies of loss during COVID-19', *Dialogues in Human Geography*, 10(2), pp 107–111.

Markham, A. N. (2021) 'Echolocation as theory of digital sociality', *Convergence*, 27(6), pp 1558–1570.

Martinuzzo, J. A. and Sangalli, H. L. J. (2019) 'O luto compartilhado no infoterritório: morte e intimidade transformadas no Facebook', *Revista ECCOM - Educação, Cultura e Comunicação*, 10(19), pp 47–62.

Navon, S. and Noy, C. (2023a) 'Conceptualizing social media sub-platforms: the case of mourning and memorialization practices on Facebook', *New Media & Society*, 25(11), pp 2898–2917.

Navon, S. and Noy, C. (2023b) 'Like, share, and remember: Facebook memorial Pages as social capital resources', *Journal of Computer-Mediated Communication*, 28(1). https://doi.org/10.1093/jcmc/zmac021

O'Connor, M. (2024) 'Grief universalism: a perennial problem pattern returning in digital grief studies?', *Social Sciences*, 13(4), p 208.

Olson, P. R. (2022) 'To bear a corpse: home funerals and epistemic cultures in US death care', in Dawdy, S. L. and Kneese, T. (eds) *New Death: Mortality and Death Care in the Twenty First Century*. School for Advanced Research Advanced Seminar Series, University of New Mexico Press, pp 219–238.

Papacharissi, Z. (2014) *Affective Publics: Sentiment, Technology, and Politics*. Oxford University Press.

Pasquali, F., Bartoletti, R. and Giannini, L. (2022) '"You're just playing the victim": online grieving and the non-use of social media in Italy', *Social Media + Society*, 8(4). https://doi.org/10.1177/20563051221138757

Perluxo, D. and Francisco, R. (2018) 'Use of Facebook in the maternal grief process: an exploratory qualitative study', *Death Studies*, 42(2), pp 79–88.

Saukko, P., Malson, H. and Brown, A. (2024) 'The ethoses of (dis)connecting with friends on social media: digital cocooning and entrepreneurial networking among people with eating disorders', *Social Media + Society*, 10(4). https://doi.org/10.1177/20563051241287284

Šiša, A. (2024) 'Ghosting on Tinder: examining disconnectivity in online dating', *Media and Communication*, 12, p 8563.

Sisto, D. (2020) *Online Afterlives: Immortality, Memory, and Grief in Digital Culture*. MIT Press.

Stage, C. (2017) *Networked Cancer: Affect, Narrative and Measurement*. Springer.

Stokes, P. (2021) *Digital Souls: A Philosophy of Online Immortality*. Bloomsbury Academic.

Stroebe, M., Schut, H. and Boerner, K. (2010) 'Continuing bonds in adaptation to bereavement: toward theoretical integration', *Clinical Psychology Review*, 30(2), pp 259–268.

Sumiala, J. (2022) *Mediated Death*. Polity Press.

Treré, E. (2021) 'Intensification, discovery and abandonment: unearthing global ecologies of dis/connection in pandemic times', *Convergence*, 27(6), pp 1663–1677.

Turner, R. B. and Stauffer, S. D. (2023) *Disenfranchised Grief: Examining Social, Cultural, and Relational Impacts*. Routledge.

Wagner, A. J. M. (2018) 'Do not click "like" when somebody has died: the role of norms for mourning practices in social media', *Social Media + Society*, 4(1). https://doi.org/10.1177/2056305117744392

Walter, T. (2017) 'How the dead survive: ancestors, immortality, memory', in Jacobsen, M. H. (ed) *Postmortal Society: Towards a Sociology of Immortality*. Routledge, pp 19–39.

Afterword: Reflections on the Role of Atmospheres in the Mediation of Everyday Life

Peter Lunt

Introduction

The case studies presented in this volume illustrate the role of emotions in the atmospheres generated by media, both by design and through the participation of audiences and the public. The studies seek to understand the emotional experiences of audiences and users of digital and social media as these arise in everyday audiences in a media-saturated world. What feelings are evoked through the mediation of family life, feminist activism, participation in secular pilgrimages and the experience of mourning? The interpretations of the emotions involved in the mediation of these social practices provide insights into what it is like to live in a media-saturated world with its expanded horizons, multiple connections, platforms and apps for everything and everything that at once invites engagement and social connection while also being managed by influential media organizations. Whereas linear media had its effects at a distance by framing ideological messages that audiences interpreted from their social positions, digital and social media are complex formations of affordances and resources woven into everyday life's practices as pervasive incursions into social routines and relationships.

AFTERWORD

In these reflections, I will review the chapters and seek to understand what we have learned about media atmospheres and the emotional experiences of audiences and participants as they go about their everyday lives. I will identify key concepts that emerge from the interpretations provided and reflect on the embedding of social and cultural practices. Although the focus is on the culture of experience, the family is a social institution responsible for socialization, reflecting social roles and norms. Also, feminist activism is a social movement standing in a critical relation to established power; secular pilgrimages are ritual forms which intersect with tourism; and mourning is an ostensibly private experience of grief and loss interacting in complex ways with social media. A central idea running through the chapters, in addition to exploring emotional aspects of the mediation of everyday life, is the idea of affective atmospheres. Böhme and Engels-Schwarzpaul (2017) have explored this idea in some depth as a philosophical analysis of the architectural atmospheres (Lunt, 2025). Böhme and Engels-Schwarzpaul (2017) analyse atmospheres as experiences of the felt body encompassing 'natural' environments and the designed spaces of architecture. Architecture, as designed spaces, nevertheless leaves open the possibility that those present in the space or place can contribute to the atmosphere through their actions and interactions and the public expression of their emotions. Think of a rock venue and a cathedral – both are designed spaces. However, the concert experience requires the voluble, visible and emotional response of the audience. Think back to public events televised during Covid to realize what was missing was the audience reaction as part of the atmosphere. Similarly, a cathedral can be experienced in private as a spiritual environment, but this changes when it is full of people having a religious experience, following the rituals of the service, as well as singing or praying. The chapters in this collection address the complex question of how the media create atmospheres by design or incorporate the felt experience of audiences and participants. The integration of media into the kinds of everyday practices explored in this book complicates these arguments further since the media seems not simply to be constituting a space of experience but also provides affordances and resources critical to everyday audiences in a mediatized world.

In their approach to the interpretation of affect and atmosphere in mediated social practices, the authors draw on insights from trends in media and communication theory, including mediatization, media practice, digital media affordances, and mobility as well as affect, to understand how people make practical sense of everyday life through their engagement with media. They coin various concepts that seek to capture the atmosphere and affect in their empirical studies. The practices studied in the book are mediated social forms. Family is both an important social institution and a constitutive practice formed through family relationships in everyday life. Feminist activism is a social movement that critically relates to established power. As a context for engagement and solidarity, secular pilgrimages are ritual forms that reflect the relation between macro processes of secularization and individualization while affording opportunities for personal experience and self-development. Mourning reflects cultural traditions, rituals and personal experiences played out partly through social media. The studies also capture several transitions in the study of media audiences in response to the changing media environment, particularly in loosening the relationship between technologies and viewing schedules in a cross-media environment. The domestication of multimedia technologies in the home and mobility extended the potential for relationships at a distance. These changes are reflected in a shift in focus in audience research from media texts and audience interpretation to media experience as media are increasingly understood as creating an environment and providing resources for social practices as part of the affective turn (Ahmed, 2014). Media is both an incursion into everyday life and provides affordances and resources for social practices such as 'doing family', participating in activism, taking care of and developing the self as a project, and rituals of mourning.

Analytics: concepts and interpretations

The chapters in the book explore the affective atmospheres generated by media and modified by the actions and interactions of audiences and the public in the context of the mediated social practices of family life, feminist activism, secular pilgrimages and mourning. The studies are influenced by assumptions drawn from critical insights developed by media researchers in theorizing the changing media environment. The transition from linear to digital media changes the affordances and resources available to audiences and users. The weaving of digital and social media into social practices extends the power of the media into everyday life contexts. It raises the challenge of evolving ways of researching and understanding how mediated experience is shaped by and constitutes atmospheres. Rather than starting with the media and examining how they create atmospheres, frame experience, patterns of reception, interpretations and influence family relations, the studies aim to understand the interaction between media-generated experiences and the contribution of audiences, participants and users to affective atmospheres and experiences. A Twitter storm, for example, is an effect of the framing of participation by the platform and by the actions of multiple participants responding to each other's posts to create a spike in activity. The studies recognize the tensions and ambivalence that this brings to the mediated practices of everyday life. The focus on affect also shifts the research away from the text-reader metaphor to embrace the bodily and affective aspects of experiencing and participating in mediated affective atmospheres. Another important contribution is moving away from focusing on audience interpretation of media content to embrace the prelinguistic, affective and social phenomenology of audience engagement with media, each through media as part of everyday social practices.

In the Introduction, the focus of the collection on mediated affect and media atmospheres is established through a discussion of various concepts used to analyse the case studies in the book, for example 'affective atmospheres', 'digital media atmospheres', 'felt experience', affect, emotion and mood. The focus is on analysing habitual, tacit, unspoken felt experiences in the context of atmospheres created by media and through the actions and interactions of participants. Key themes in analysing how atmospheres affect everyday social practices include variations in the intensity of feelings and the ambivalences and tensions that arise from emotional engagement and immersion in media atmospheres. Several analytic distinctions are identified. For example, there is a contrast between the role of the media atmosphere routine and mundane social practices and the role of the atmosphere in disruption.

Chapter 1, 'Post-Digital Parenting: The Relational-Affective Network of the Family' by Francisca Porfírio, Ana Jorge and Rita Grácio, explores the interpolation of the transformation of the home and family relationships arising from the establishment of the mediatized home. How are relationships and practices in the post-digital family, which embraces digital media as part of family life, in the practices of 'doing family' as a mundane reality? Constituting and sustaining family through relationships, routines and interactional encounters now implicate digital media, and the chapter examines the affective atmospheres or emotional climate (Lunt and Stenner, 2005) within the family. While social media can be a resource for sustaining family routines and relations, parents express concerns about their children's digital health, screen time and potential exposure to risky and harmful content and online relationships. Regarding family life, parents also have concerns about how immersion in digital media creates social and role distance, disenchantment and scepticism (see Livingstone and Sefton-Green, 2016). The family illustrates the importance of mediated sociality, interaction intersubjectivity and the experience of being in the social world together. Nor is the impact of digital affective climates limited to the family home. The study of mediated family life involves complex layers of digitization that connect the home, school, social and cultural activities and friends (Livingstone and Sefton-Green, 2016). In the face of these challenges, parents seek to encourage positive engagements within the family through digital media by embracing digital family practices, sharing in media events and using digital media to cultivate family relationships in, for example, WhatsApp groups. Nevertheless, the parents in the study found their children's online activities challenging and a source of family conflict and expressed ambivalence towards technical solutions such as parental controls. Mothers shoulder the burden of the emotional labour surrounding the role of digital media in family life, recognizing the intensification of emotions and the ambiguities of parenting roles in managing mediated family life.

The post-digital theme in the study of family life is also reflected in the chapter that explores the augmentation of feminist activist practices through digital and social media – creating an atmosphere in which identity work, which in this context is digital by default, plays an important role in identity and political engagement. The study focuses on the activities surrounding International Women's Day in March 2023 in Lisbon. In Portugal, feminism is a state-sponsored activity rather than a grassroots life-political movement in which context the affordances of digital and social media create opportunities for engagement. The chapter documents the creation of a 'feminist filter bubble' as a safe space to the side of the official national sponsorship of the march. The affective aspects of the emergent social media space are described as a 'purple wave' that creates an ambience that carries participants along through fleeting vibes or transient moods of varying intensity. In addition, platforms allow information sharing, advocacy and movement building through digital communities of practice. The experience and emotions of these engagements can disrupt everyday life and involve experiences of anxiety related to toxicity and online hate. Participants discuss a variety of tactics for managing intensity through small acts of disengagement and avoidance of aggression and hostility. Interestingly, there were reflections on online civic responsibility as a counter to these negative currents of feeling and ambivalence arising from the contrast between the desire to get involved and the need to manage the intensity of feelings and negotiate the experience of social media. In this, there is a tentative mapping of modalities of affective engagement and different social media platforms, with Facebook and Twitter more overtly politicized as vehicles for news, opinion and discussion. In contrast, primarily through stories, Instagram created a sense of fun and was aesthetically pleasing as a kind of 'peaceful bubble' accompanying events. This mixture of transient moods, pervasive pressures and civic responsibility affords a variety of atmospheric assemblages accompanying the march. The positive and negative or banal affective states variously energized people, creating feelings of despondency, negative affect and general pessimism. Participants experienced these different moods with varying intensity through their engagement, requiring mood management to negotiate the dynamics of feeling and surges or shifts in moods and feelings, which are sometimes invigorating and sometimes overwhelming or provoking feelings of anxiety. These mixed feelings and varying intensities of involvement, along with a proliferation of negative posts and a sense of a toxic technoculture that normalizes controversy, hate and polarization, lead to alienation and exhaustion of feeling burnt out in tensions. Interestingly, Chapter 2 also included a discussion on 'response-ability' to engage, immerse and constitute a caring environment with positive energy (Chen and Lunt, 2021).

Chapter 3 by Ana Jorge, Filipa Neto, Ana Kubrusly and Edna Santos explores the 'affective temporality' of the secular pilgrimages to Santiago de Compostela, Galicia Sanctuary Fátima and participation in the World Youth Day held in Lisbon in 2023. There are extensive online resources related to these activities in preparation and anticipation of the pilgrimages, during the journey and event, and later memories and reflections on the experience. Participants engage in the full panoply of digital and social media, including maps, apps, repositories, Instagram, Facebook groups and pages, and TikTok. Participant observation data was collected through 30 interviews, revealing affective and logistical anticipatory work, a 'premeditation that mobilises affect' (Nikunen, 2023, p 178), as the interweaving of planning and feeling – temporal and affective structuring of intimacies, socialities and relationships. The analysis starts from the paradox of societal secularization, the relative decline of formal religious observance, and the increasing interest in spirituality through meditation, holistic wellness, or yoga. Pilgrimages represent an orientation towards tradition and an escape from the pressures of everyday life in modernity, a sacred to the profane of everyday life, a quest rather than an unfolding set of activities. Pilgrimages are structured in time and place as a journey which lends itself to narrative form supported by a plethora of digital and social media as an unfolding quest narrative. A further dimension of pilgrimage is focusing on self-improvement or development as part of the self (Giddens, 1991). There is a substantial overlap with forms of tourism that are similar structures, such as cruises, skiing and walking holidays, monastic retreats and spa breaks. The smartphone emerges as a critical technology in pilgrimage because of its multiple functions and intersection with smartwatches, GoPro and cameras, which are all used to navigate and document the journey, capture relevant information, share the experience and connect to others through social media. The ritual journey involves a series of experiences in different stages involving various atmospheres and emotions with varying forms of attunement to shared feelings and the atmospheres of spaces. It is akin to the Stations of the Cross in Jerusalem. As in any quest narrative, various challenges must be managed, including overcrowding, the commercialization of the pilgrimages and potential overflows of emotion.

Chapter 4, by Ionara Silva, Ana Jorge and Filipa Neto, explores the affective intensities of dis/connection in mourning. The authors document how digital media reconfigures death and mourning. Through digital traces, the deceased live on, mourning persists, and bereavement is a process that is negotiated over time, balancing attempting to 'keep hold' and 'let go'. Digital technology allows space for grief, which is also governed by personal and social relationships and cultural contexts shaping how people deal with loss. The tension between the private experience of mourning and the ritualized formal processes of mourning alongside the acknowledged psychological stages of mourning all now have to be managed in the context of the blurring of the boundaries of public and private on digital media, which are reconfiguring the relationship between death, mourning and memory. The affective intensities experienced in bereavement permeate digital connection/disconnection, and mourners face complex emotional experiences as they configure their digital media use during mourning and cope with the existential and relational dimensions of loss. In this, they face nuanced decisions of what to make public, whether and where to 'publicize' their loss and experiences, facing potential risks of the erosion of privacy, loss of control over grieving and memory, and the inauthentic sharing of emotions. Some participants were overwhelmed by social media responses, and others withdrew from social media and could not work out how to express their feelings of mourning. How does one 'share' emotions recording loss of intimacy and grief on social media? These thoughts suggest that emotional norms are prevalent on social media related to keeping things light (Chen and Lunt, 2021). Against this, there is the potential for support through posting news of bereavement on online safe spaces and support groups that can be useful by providing reciprocity and empathy. The chapter concludes: 'Mourning is a complex field where affect, social norms and platforms converge.'

Reflections: media practice, emotion and power

The chapters in this volume provide insights into the emotional aspects of digital and social media, such as felt atmospheres that shape social practices and offer affordances and resources for shared experiences, interaction and intersubjectivity. In the context of assumptions about the changing media environment, the various studies of emotional engagement in everyday audiences mediated by digital and social media capture tensions, contradictions, variations in intensity and commitments. In this final section, I will link these studies and concepts to various ways in which affect has been studied in media and cultural studies to put the findings presented here in context and tease out some of the theoretical questions. I refer to Chouliaraki's (2006) analysis of the spectatorship of suffering, Papacharissi's (2014) analysis of affective publics and Banet-Weiser's (2018) analysis of popular feminism as a point of contrast with the chapter on feminist activism. These are supplemented by Couldry's (2004) foundational work on media practice and Burgess and Baym's (2020) analysis of Twitter as a combination of cultural artefact and communication infrastructure. I will end by recognizing the sociological aspects of the case studies presented here: the family as a social institution and a constituted practice; feminist activism as a social movement; the ritual of secular pilgrimage and its development as tourism; and the norms and rituals of mourning. To this effect, I will provide reflections on Ahmed's (2014) analysis of the cultural politics of emotion to address the central conundrum that sits behind these studies: the relationship between the power of the media to structure affective atmospheres, emotional engagement and the constitutive practice of participants in their engagement with their mediated felt experiences.

Chouliaraki's (2006) analysis of the spectatorship of suffering takes us back to linear media and journalism through an analysis of the power of the media to shape feelings towards distant suffering in the context of war, famine and natural disasters in ways that influence both feelings and our subject positions as moral agents in taking up positions and actions towards what we observe. Chouliaraki mapped genres of representations of distant suffering onto associated feelings shaped by viewing images and stories of distant suffering, which she argued positions viewers as social actors, passive observers, volunteers or donators, or distant moral agents expressing sympathy. From Chouliaraki's work, the critical questions are structuring experience through the genre that evokes feelings associated with our potential for engagement and agency. Although the idea of genre is linked to the text/reader metaphor and is less directly relevant to questions of the role of affect in orientation to media practices in everyday life. Chouliaraki's analysis challenges us to find patterns in the accounts given in this book of the relationship between mediation, feelings and action. For example, how are technological affordances linked to feeling patterns and actions? We might also link ritual forms of pilgrimage to feelings of immersion in space and place in support of an action-oriented quest related to self-development.

Papacharissi's (2014) analysis of affective publics takes up some of the themes of Chouliaraki's analysis of how storytelling on social media shapes shared feelings among users and is connected to the foundation of affectively charged political engagement. These ideas influence our views of the meanings of the public sphere in the digital age, the emergence of structures of feeling through shared responses on social media and how these are reflected in the way that news is oriented towards feelings as 'affective news', a blend of feelings and facts that resonate with emerging currents of feeling. Papacharissi recognizes that this combination of elements leads to forms of public connection and movements that are not grounded in traditional conceptions of organized social movements and political commitments. These ideas resonate with the findings in this book, which show that the affective publics of feminist activism surrounding International Women's Day and secular pilgrimages are losing affiliations and connections. Both studies emphasize personal development and the project of the self (Giddens, 1991), complementing Papacharissi's account of resonance as the link between media and the affective public. The 'fit' between media and social practices affords processes of self-reflection and development and an emergent public life.

The chapters in this book emphasize the complex interleaving of digital and social media and everyday audiences. Each case study approaches its topic as a social practice and seeks to capture how affective atmospheres and feelings are implicated in mediated social practices. Engaging with feminist activism, constituting family life, participating in a secular pilgrimage and managing social media during mourning all illustrate the interleaving of affect, media and social practice. Couldry (2004) recognized the importance of a focus on mediated social practices as a way of going beyond the focus on audiences as interpreters of texts and the focus on the structuring properties of media and questions of what it is like to live in a media-saturated world (Ang, 1996) to affirm the importance of media power by asking what it is like to live in a society dominated by powerful media. These ideas focus on acknowledging how everyday life is saturated in media and that audiences must develop strategies for dealing with media and the intrusion of media and market logic into everyday life. Living in media provides affordances and resources that are taken up in everyday audiences while also involving a confrontation with difference, conflict and power. These themes complement and potentially extend the treatment of engagement with feminist activism, the commercialization of pilgrimages and the contradictions of social media engagement and family life and social media identified in the case studies presented here.

There are also interesting resonances between these case studies of atmospheres and felt experiences and Banet-Weiser's (2018) analysis of popular feminism, which also points towards populism in media engagement and affect. The contradictions revealed in the feelings of being overwhelmed and facing potential hostility and conflict in social media in all the case studies reflect the paradox that social media empowers both popular feminism and misogyny. Popular feminism fits the forms of articulation in social media, such as slogans, hashtags and commercialization in advertising and celebrity culture. However, the availability of these representations and engagement also provides resources for widespread misogyny, reflecting broader processes of branding and potentially diluting what Banet-Weiser calls the structural critique of patriarchy and capitalism. Popular feminism involves fragile visibility and ambivalence, trends reflected in the accounts of both mediated family practices and the ambivalence and contradictions of engagements with feminist activism.

The interpretations of these studies of affective atmospheres and felt experiences resonate with Ahmed's (2014) analysis of the affective turn and the cultural politics of emotion. First, recognizing the affective turn is not about understanding emotions as internal psychological states or feelings but that emotions are social, fluid and implicated in the circulation of meanings, objects, collectives and individuals. As such, emotions are socially and culturally located and confront power in their articulation in social interactions, relationships and practices (Lunt, 2025). Second, Ahmed (2014) also analyses the stickiness of emotions afforded to public figures and media circulating through social life. In contrast, the ambivalences and distancing from media identified in this book point to how negative emotions can mark differences and risks. However, emotions are critical to belonging and attachment to collectives. The engagement of families in the chapter on the mediated constitution of family life also reflects Ahmed's (2010) account of the family as the repositioning of firmly held feelings and expectations of happiness, which, as identified here, can be hindered by the incursion of digital and social media into everyday life.

Finally, what have we learned about the affective turn, affective atmospheres and felt experiences from the studies presented in this book? The book and my commentary here have focused on the aesthetics of mediated felt atmospheres, the different forms that mediation gives to experience, and the strategies and tactics adopted by audiences and the public. However, Böhme and Engels-Schwarzpaul (2017), in their influential analysis of architecture as felt atmospheres, combined an analysis of the forms of built environments as designed spaces defined in terms of their physical properties such as shape, proportion, light and sound. In addition, they put equal weight on the actions and interactions of those present and their voices, glances, movements and vocalizations. Felt atmospheres are an amalgam of material conditions and participants' activities that potentially create the grip that atmospheres induce. Bohme and Engels-Schwarzpaul (2017) used various metaphors to capture this amalgam of material conditions and dispositions, including the idea of atmospheres as 'tuned' spaces, implying that an account of intersubjectivity and the social phenomenology of mediated felt experiences is required to understand media as affective atmospheres. We get numerous glimpses of experience in the analyses presented in this book that suggest that the study of media atmospheres is a promising area of development that carries the affective turn into the study of the interweaving of digital and social media with everyday audiences, with implications for understanding participation and committed engagement in the politics of everyday life.

References

Ahmed, S. (2010) *The Promise of Happiness*. Duke University Press.
Ahmed, S. (2014) *The Cultural Politics of Emotion*, 2nd edn. Edinburgh University Press.

Ang, I. (1996) *Living Room Wars*. Routledge.

Banet-Weiser, S. (2018) *Empowered: Popular Feminism and Popular Misogyny*. Duke University Press.

Böhme, G. and Engels-Schwarzpaul, A. (2017) *Atmospheric Architectures*. Bloomsbury Academic.

Burgess, J. and Baym, N. K. (2020) *Twitter: A Biography*. New York University Press.

Chen, S. and Lunt, P. (2021) *Chinese Social Media: Face, Sociality, and Civility*. Emerald.

Chouliaraki, L. (2006) *The Spectatorship of Suffering*. Sage Publications.

Couldry, N. (2004) 'Theorising Media as Practice,' *Social Semiotics*, 14(2), pp 115–132.

Giddens, A. (1991) *Modernity and Self-identity: Self and Society in the Modern Age*. Polity.

Livingstone, S. and Sefton-Green, J. (2016) *The Class: Living and Learning in the Digital Age*. New York University Press.

Lunt, P. (2025) 'The felt experience of the atmosphere: implications for audience research', in Hill, A. and Lunt, P. (eds) *The Routledge Companion to Media Audiences*. Routledge, pp 471–482.

Lunt, P. K. and Stenner, P. (2005) 'The Jerry Springer Show is an emotional public sphere.' *Media, Culture & Society*, 27(1), pp 59–81.

Nikunen, K. (2023) 'Affective temporalities of digital hate cultures', in Lünenborg, M. and Röttger-Rössler, B. (eds) *Affective Formation of Publics*. Routledge, pp 173–190.

Papacharissi, Z. (2014) *Affective Publics: Sentiment, Technology, and Politics*. Oxford Academic.

Index

References to tables appear in **bold** type.

A

#ACaminhoDo8M 67
actions and interactions 155
activists 67, 75
addictiveness 2
affect 7–9, 15, 145–146
 see also emotions
affective atmospheres 10, 11–12, 58, 143, 145–146, 154–155
affective ductility 42–43
affective engagements 13, 147
affective labour 32, 40–45, 48–49, 75
affective negotiations 44
affective publics 152
affective relations 32, 48
affective temporalities (Nikunen) 90
affective turn 7, 154–155
affect theory 7–8
Ahmed, S. 151, 154
albergue hosts 100
algorithmic mediation 135–136
algorithmic structures 71
Almeida family 35, 37
Almeida, Clara (9 year old daughter) 38, 39, 42–43
Almeida, Hugo (44 year old parent) 38
Almeida, Simão (16 year old son) 42–43
'always-on' culture 2, 89
ambience 58–59
ambivalence 3, 12, 40, 74, 122, 145, 147, 148, 154
Andelsman, V. 47
Anderson, Ben 10
anticipation 90, 95–98
anxieties 1
architecture 143
Association of Internet Researchers (AoIR) 16, 35, 95, 119
atmospheres 3, 10–13, 144, 145–146, 155
audience engagement 20, 145–146
audience research 9, 11
avoiding notifications 126–127

B

Banet-Weiser, S. 151, 154
Barroso family 35, 40
Barroso, Ana (mother) 36, 38, 40–41, 45, 47–48
Barroso, Margarida (11 years old daughter) 38, 40, 41
Baym, N. K. 151
Bengtsson, S. 74
bereavement
 disconnecting from social media 122–123
 emotions 124
 feelings of isolation 130–131
 interviews 117–121, **120**
 'personal mourning logs' 127
 therapeutic practices 125
 'toxic feelings' 123–124
 unsolicited notifications 126
 validating feelings 127
 see also grief and grieving; mourning
bereavement support groups 119, 124–125, 150–151
Berlant, L. 89, 131
Böhme, G. 10, 143, 155

Brazil 117, 118, 132–133
Burgess, J. 151

C

Caldeira, S. P. 77
Caminhos de Fátima 92
  see also Sanctuary of Fátima
Camino de Santiago 91, 93, 97, 100, 102
  see also Santiago de Compostela pilgrimages
Carvalho family 35
Carvalho, Diana (infant) 43, 46
Carvalho, Graça (mother) 36, 43, 44–45, 46
Carvalho, João (father) 38, 43, 46
cathedrals 143
children 30–49
  consenting to sharenting 39
  digital health 37–38
  digital media at meals 33, 40
  digital media in restaurants 37
  instant messaging 47
  and intensive parenting 45–46
  interactions between siblings 42–43
  intimate surveillance 46
  introducing to other interests 36–37
  joy and happiness 48
  managing technology at home 41–42
  pacifying and entertaining 37, 49
  stretch the limit 42
  study of family life 32–36, *34*
  tensions and disagreements 40–45
  see also families; parents
Chouliaraki, L. 151, 152
Christian pilgrimages 91
  see also pilgrimages
coding processes 66
communal support 131–134
communicative *genres* (Lomborg) 69
communitas 89, 97, 100, 103, 107–108, 131

concert experiences 143
conflicting desires 12
connection and disconnection 16
contemporary feminisms 77
continuous surveillance 41
Couldry, N. 99, 151, 153
COVID-19 pandemic 16, 118–119, 125
cultural hierarchies 37
cultural norms in grief 133–134
cultural studies 7

D

daily rituals and digital media 2–3
Days in the Dioceses (WYD 'pre-events') 98
death 115–116, 150
  see also bereavement; grief and grieving; memorialization
definitive disconnection 4
della Dora, V. 88–89, 92
Democratic Republic of the Congo (DRC) 6
detachment and self-protection 126–127, 135
detox camps 5
development 149, 153
Dias family 35, 41–42
Dias, Ângelo (5 years old) 42
Dias, Daniel (16 years old) 40
Dias, Dulce (stepmother) 41–42, 45–46
Dias, Guilherme (12 years old) 41
Dias, Júlio (father) 41–42
Dias, Lisa (11 years old) 42
digital activism 58, 77
'digital by default' logic 77
'digital detox' 5
digital devices *see* smartphones
'digital dualism' 58
digital health 37–38
digital hyperconnectivity 16
digital labour 32, 45–47, 59
digital media
  and affective atmospheres 12
  atmospheres 12, 18, 90, 146

INDEX

daily rituals 2–3
interleaving with everyday audiences 153
pilgrimages 89–90
regulating children's emotions 37–38
saving/stealing time 5
social practices 145
wayfarers for pilgrims 99, 107
digital memories 104–105
digital parenting 30–31
dis/connection 3–4, 135
disconnection 4–6, 77–78, 116, 122–123
'disenfranchised mourning' 117–118
disentangling from digital media 3
Disney+ 40
'doing family' 31, 32, 36–40
domestication of multimedia technology 144
doulas 136

E

effervescence 89
Ehn, B. 11
emotional labour 44, 147
emotions
and affective turn 154
duality 146
and moods 31, 146
socialization 8
see also affect
Engels-Schwarzpaul, A. 143, 155
Esteves family 35, 37
Esteves, André (8 years old) 39, 40
Esteves, Dulce 47
Esteves, João (father) 36
Esteves, Lina (mother) 36, 42
Ethical Guidelines (AoIR) 35, 95, 119
ethnographic studies 33–36
everyday life 9–10
see also offline activities
excitement of living, and sharing 102

F

Facebook
algorithmic structure 71
outdated platform 70–71
politicized for news and opinion 148
Facebook groups
bereavement pages 118–119, 127
death of a child 133
Fátima pilgrims **94**, 97, 101, 103–104, 106–107
mainstreaming funeral details 121–122
messages 128
multimodal analyses 118
phatic communication 133
Santiago de Compostela pilgrims **94**
semi-public environments 131
shared consciousness for bereaved 132
WYD **94**, 106
face-to-face grief support groups 132
families
as a constitutive practice 144
expectations of happiness 154–155
and Instagram 39
and media 30–32
mediated social form 144, 146–147
offline activities 36, 38
post-digital themes 147
socioconstructivist approaches 31
see also children
Family Link app 40, 45
family meals 33, 40
Fast, K. 31
'features for disconnectivity' 6
feelings of community 68–69, 75
felt atmospheres 146, 155
felt experiences 154–155
feminism
characteristics and communities 70
'digital by default' 77

as 'labour of passion' 75
Portugal 147
and social media ambiences 58
without social media 66
#FeminismoPortugal 61
feminist activism 58, 75, 143, 144, 153
feminist communities 57, 68
feminist content 73–74
feminist filter bubbles 69, 71, 147
feminist potentialities 57–58
fieldwork 16

G

gendered norms in grief 133–134
genres 152
'ghosting' 128
Gibbs, A. 7
GIFs 96–97, 100
Giphy **94**, 100
Global South 133–134, 136
'good parenting' 37, 48
Google Classroom 37
Google Maps **94**, 104
Google searches 124
grief and grieving 117, 121–125
 as a communal experience 124
 communicating 133
 cultural and gendered norms 133–134
 reading about 125
 representations of 115–116
 and sharing emotions 150–151
 suspending social media 135
 see also bereavement; mourning
grief communities 131–133
group chats with strangers 40–41

H

Hermes, J. 7–8, 11
Hypatiamat (educational website) 37

I

Iberian Peninsula 1–2
individualization and secular pilgrimages 144

Instagram
 8M march 60–61, 67, 68, 72
 ambivalent platform 71–72
 analyses 35
 cultural hierarchies 37
 decompress and distancing from 76
 family activities 39
 Fátima pilgrims 93, **94**, 101
 judgemental posts 44–45
 less toxic environment 72
 parents **34**
 'peaceful bubble' 148
 Santiago de Compostela pilgrims 93, **94**, 97, 100, 103
 shared 'digital memory spaces' 105
 sharenting 46
 social support 43–44
 study 32–33
 WYD 93, **94**, 105
instant messaging 47
Insta-Save 33
intensive parenting 45
International Women's Day march (8M march) 147
 feminist activism 153
 Instagram 60–61, 67, 68, 72
 interviews 61–66, **62–65**
 offline feminist spaces 79
interviews 15
'intimate publics' and reciprocity 132
intimate surveillance 46

J

Johansson, S. 74
Jorge, A. 4, 47

K

Kaun, A. 127
Keller, J. 70
kindergarten apps 46
Kinshasa, DRC 6
Koivunen, A. 95–96

INDEX

L

'labour of passion' 75
Leaver, T. 46
Lefebvre, H. 90, 107
Lim, S. S. 31, 45
linear media 142, 145, 152
Lisbon 61, 92, 93, 98, 147
livestreams 100
Lomborg, S. 4, 69
Lúcia, Irmã 92
Lunt, P. 10
Lupinacci, L. 78

M

Maddrell, A. 88–89, 92
Madsen, O. J. 4
managing connectivity 6–7
Mannell, K. 42
Markham, A. N. 68, 116
Markham, Tim 12
materiality and intersubjectivity 10
MaxQDA software 35, 66, 95
mealtimes 33, 40
media
  audience studies 9
  in everyday life 30, 144–145
  and social practices 153
media atmospheres 143, 146, 155
media power 153
media-saturated world 142, 153
mediated affective
  atmospheres 145
mediated family life 146–147
mediated social forms 144
mediated social practices 153
memes 101
memorialization 127, 129–130
  see also death
memorialization cultures 136
mental health crises 2
migrants 136
miscarriages 123, 130
modulating connectivity 6
Moe, H. 4
monitoring digital activities 40–41

moods 8, 74, 146
mothers 130
mourning 115–137
  accessing social media 126
  complexities 134–135
  COVID-19 pandemic 125
  cultural traditions 144
  'disenfranchised mourning' 117–118
  interviews 118–121, **120**
  notifications of
   memories 126–127
  psychology models 134
  sharing on social media 122
  vulnerability 116
  see also bereavement; grief
   and grieving
Musk, Elon 70

N

Navon, S. 118, 127
negative emotions 154
Nehring, D. 88
Nikunen, K. 90, 95
Nilsson, M. 91
Nintendo Switch 38
non-mediated activities 39
non-participant observations 118
nostalgia 105, 106
Noy, C. 118, 127

O

offline activities
  families 36, 38, 39
  social movements 79
  support groups 136
offline feminist
  mobilizations 60–61
*On&Off: atmospheres of
  dis/connection* project 13
Once Upon a Time (Disney+) 40
online communications 5
online memorialization
  practices 127–129
'online mourning guestbook'
 (Navon and Noy) 127

online semi-public environments 131
organized social movements 153
oversharing 19, 129

P

Paasonen, S. 2
Papacharissi, Z. 8, 151, 152–153
parental bereavement 118
parental sharing 38–39
parent-child interactions 39
parents
  anxieties 47–48
  conflicts with children 38, 42
  continuous labour 48–49
  cultural hierarchies 37
  digital media and feelings of assurance 48
  dissonance over sharenting 43
  domestication theory 30
  emotions with digital media 48
  on Instagram **34**
  monitoring digital activities 40–41
  relational labour 45, 47, 49
  see also children; sharenting
Passo-a-Rezar app 98
path and *journey* metaphors 88
'peaceful bubbles' 148
personal and professional spheres 2
'personal mourning logs' 127
pervasive pressures 73–79
PhantomBuster 61
phatic communications 133
photographs 101
pilgrimages 88–108
  advice to pilgrims 97
  as an alternative economy 89–90
  'always-on' culture 89
  anticipation 90, 95–98
  Christian 91
  defining 88
  divergent portrayals 105–106
  effervescence 89
  memory-making 104
  navigational information 99
  path and *journey* metaphors 88
  as a 'pause' in everyday life 99
  pilgrims celebrating experiences 103
  post-secular forms 108
  preparing for 95–96
  presence 90, 99–103
  (pro)longing 90, 103–107
  as a quest 149
  relational and social dimensions 89
  religious and spiritual crescendo 98
  ritual dimension 102–103
  secular 144
  self-improvement 149
  smartphone 149
  social media as a diary 101
  social media platforms 92–93
  temporally driven collective feelings 90
  and tradition 149
  transformative experiences 103
  transnational links 14
PineTool 33
platformized feminisms 18, 58, 70, 79–80
platform *vibes* 59
political responsibilities 75
popular feminism 154
Portugal 13–14, 32
  8M march 60
  bereavement study 117–121
  ethnographic studies 32–36
  feminism 147
  memory cultures 134
positive emotions 48
positive experiences 69, 106
post-digital parenting 31–32, 48, 147
posting instantaneously 101–102
posting pictures
  and memorialization 129
  pregnancies and ultrasound scans 123
power outages 1–2
presence 90, 99–103
presencing (Couldry) 99

INDEX

(pro)longing 90, 103–107
public spaces 37
'purple wave' 147
Pype, K. 6

R

radical disconnection 4
radios 1
reciprocity 132
relational labour 45, 47, 49
religious experiences 143
research
  disconnecting from social media 4
  families and media 30
research ethics 16
ritual journeys 149
Rota Carmelita 92
routines and rituals 9

S

Sanctuary of Fátima 89, 99, 149
  biannual celebrations 91–92
  digital media corpus **94**
  Facebook groups 97, 101, 103–104, 106–107
  religious and spiritual crescendo 98
Santiago de Compostela pilgrimages 89, 91, 149
  digital media corpus **94**
  GIFs 96–97
  Instagram posts 97, 101, 103
  see also Camino de Santiago
Schmitz, H. 10
'screen time' functions 6
secular pilgrimages 144
  see also pilgrimages
selfies 101
self-protection 126–127, 135
self-reflection and development 153
sharenting 31, 38, 39, 43, 44, 46, 48–49
  see also parents
sharing
  'digital memory spaces' 105

and excitement of living 102
  mourning 122
  offline family activities 38
  ritual dimension 102–103
  simultaneous translations 100–101
Šiša, A. 116
slacktivism 58
small acts of engagement' (Picone et al, 2019) 77
smartphones 67, 99–100, 125–126, 149
social echolocation theory (Markham) 68
social inequalities 5–6
social media
  addictiveness 2
  democratizing potential 6
  feminist organizing 66–67
  interleaving with everyday audiences 153
  memorializing the dead 127
  mood management 74
  platform *vibes* 59
  political uses 76–77
  providing *reassurance* 68
  representations of death and grief 115–116
  scrolling being 'toxic' 123
  social practices 145
  social support 43–44
  symbolic spaces of interaction 69
  taken-for-granted presence 9
  toxic technoculture 74–75
  unwanted scrutiny 44–45
social media ambiences
  and affective states 69
  and feminisms 58
  shaping experiences 79–80
  transient moods 18
social practices 144–145
social support 43–44
societal secularization 149
socioeconomic privilege 136
Sørensen, A.S. 69
Sousa, Marcelo Rebelo de 101
spectatorship of suffering 152

Stations of the Cross 149–150
Stewart, K. 12
Stiernstedt, F. 127
storytelling 152
stress and anxieties 48, 76
suicide 130, 134
support groups 131–133, 136
Syvertsen, T. 5

T

talking on camera 36
Tesfahuney, M. 91
textures (Sørensen) 69
TikTok 40, 43, 93, **94**, 96, 97–98, 105
'toxic feelings' 123–124
toxic technoculture 74–75
'transcendent parenting' (Lim) 31, 45, 48–49
transient moods 18, 73–79, 147, 148
Turner, E.L.B. 89
Turner, V.W. 89
Twitter 70, 78, 148, 151
Twitter storms 145

U

Usher, B. 71

V

Van Der Wal, A. 74
vibes 12, 59, 69, 80, 147

W

Wahl-Jorgensen, K. 8
Wetherell, M. 9
WhatsApp groups 40–41, 133
women
  experiences in mourning 136
  Facebook groups on maternal pain 133–134
World Youth Day (WYD, Lisbon 2023) 91, 92, 149
  group photos 105
  Instagram 93, **94**, 105
  pilgrims from Venezuela 101
  'pre-events' 98
  preparing for 96
  religious and spiritual celebration 98, 99
  shared 'digital memory spaces' 105
  TikTok 97–98
  transformative experiences 103–106
  urgency to post 101–102
World Youth Day (WYD, Seoul 2027) 106

X

X (Twitter) *see* Twitter

Y

Ytre-Arne, B. 4

www.ingramcontent.com/pod-product-compliance
Lightning Source LLC
Chambersburg PA
CBHW071709020426
42333CB00017B/2197